AF598269

Hopper and Silo Discharge

Papers presented at a one-day seminar *Hopper and Silo Discharge: Successful Solutions* held at IMechE Headquarters, London, UK, on 27 November 1998.

IMechE
Seminar Publication

Hopper and Silo Discharge
Successful Solutions

Organized by the Bulk Handling Committee of
the Institution of Mechanical Engineers (IMechE)

Co-sponsored by

Institution of Chemical Engineers (IChemE)
Materials Handling Engineers Association (MHEA)
Solids Handling and Processing Association (SHAPA)

IMechE Seminar Publication 1999–4

Professional Engineering Publishing

Published by Professional Engineering Publishing Limited for the Institution of Mechanical Engineers, Bury St Edmunds and London, UK.

First Published 1999

ISSN 1357–9193
ISBN 1 86058 192 7

A CIP catalogue record for this book is available from the British Library.

Printed by The Book Company, Ipswich, Suffolk, UK.

Contents

Related Titles of Interest

Title	Editor/Author	ISBN
Handbook of Mechanical Works Inspection	Clifford Matthews	1 86058 047 5
IMechE Engineer's Data Book	Clifford Matthews	1 86058 175 7
Process Machinery – Safety and Reliability	William Wong	1 86058 046 7
Design, Selection, and Operation of Refrigerator and Heat Pump Compressors	IMechE Seminar 1998–15	1 86058 159 5
Plant Monitoring and Maintenance Routines	IMechE Seminar 1998–2	1 86058 087 4
Bulk Materials Handling	IMechE Seminar 1997–2	1 86058 106 4
Protecting the Environment – Controlling Leakages in the Process Industry	BHR Group Publication No. 30	1 86058 138 2

For the full range of titles published by Professional Engineering Publishing contact:

Sales Department
Professional Engineering Publishing Limited
Northgate Avenue
Bury St Edmunds
Suffolk
IP32 6BW
UK

Tel: 01284 724384
Fax: 01284 718692

S604/002/98

Using air cannons to solve silo flow problems

J BURKHART
Martin Engineering, Illinois, USA
G OTTOSSON
Martin Engineering GmbH, Walluf, Germany

For many years storing material in bulk has created a problem: how to get the material out of the storage vessel. Typical blockage problems include bridging, rat holing, clinging, and arching. When a blockage exists, production cannot continue until it is removed. One of the first things an operator will do is try to dislodge the material in some manner. Historically, water and air lances, silo clean-out equipment, shotgun blasts, and even dynamite have been used to eliminate material blockage. However, for many years, the most common means of removing blocked material was the use of a sledgehammer.

These storage flow problems continued for decades until new techniques were developed to reduce their occurrence and address storage flow problems before blockages occurred. Such solutions include special bin designs, vibrating bin bottoms, bin wall liners, vibrators and air cannons, just to name a few.

The following discussion will focus on the use of air cannons to prevent storage flow problems. In this discussion, we will review different styles of air cannon designs and learn basic guidelines for their selection, sizing, and placement, as well as some of the accessories, which are required to construct a complete flow-aid system. Also, examples and case histories will demonstrate how air cannons can be used to solve storage flow problems.

1. THE AIR CANNON SOLUTION

Air cannons continue to be a popular choice as a solution to flow problems because they have proven to be a dependable, durable, low-energy consumption, low-maintenance method to enhance material flow and improve productivity in industrial settings.

Some storage vessel applications, due to either their physical size or the material flow problem, require air cannons to remove bridging, rat holing, arching, or clinging material. Materials that are porous (e.g., wood chips, garbage) or that possess high cohesive strength (e.g., frozen coal, gypsum) and tend to absorb vibration do not react well to vibrators.

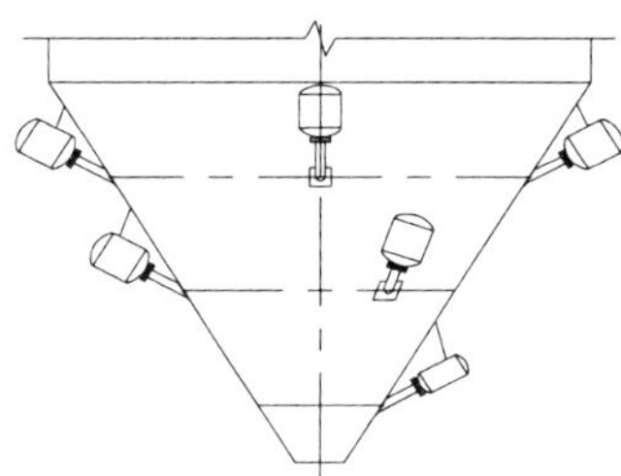

Figure 1. Typical Air Cannon System

The purpose of an air cannon is to instantaneously expel an explosive charge of compressed air or nitrogen directly into the critical areas within the storage vessel where the stored bulk material tends to clog. This explosive blast of energy breaks up the static friction and the cohesion that causes the material to arch, bridge, cling, or rat hole and allows the return of gravity flow.

Air cannon systems usually consist of a number of air cannons mounted at different levels around a storage vessel (Figure 1). The cannons are then discharged in sequence from the lowest to highest level to cover all sections of the vessel.

1.1 Typical Cannon Types

The basic air cannon consists of a storage tank and a valve assembly to expel the air rapidly from the storage tank into the material. Air cannon tanks can be acquired in various diameters and lengths to supply the proper amount of air volume and material dilation. It is important to make sure that the tank is manufactured to local Pressure Vessel Codes to ensure plant safety and government compliance.

There are two basic types of air cannon valves: the external valve and the internal valve. The external-valve air cannon is so named because the piston chamber is located outside the tank as shown in Figure 2. The valve body is constructed from honed steel or aluminum. The components inside the valve body include a plastic or aluminum piston, which slides within the

Teflon or nickel-plated inner diameter of the valve body. A piston seat provides a surface to seal the piston at pressures as low as 40 lbf/in2 (2.7 bar).

The internal-valve air cannon differs from the external-valve cannons in that the piston and valve body is located inside the storage tank (Figure 3). The components of the internal-valve air cannon include a rubber, spring-supported fill hose that attaches the air supply inlet and

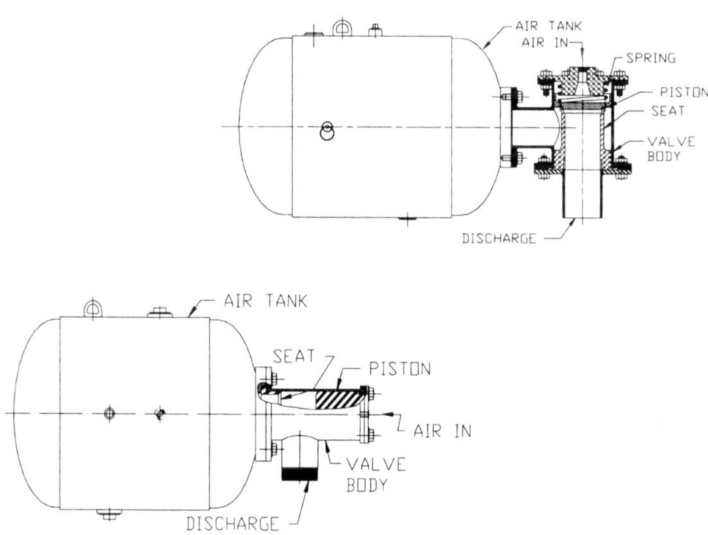

Figure2. External Air Cannons

aluminum cylinder. A check valve in the top of the cylinder allows air to flow only from the cylinder to the tank. The only moving part inside the cannon is the piston, which shuttles back and forth in the cylinder on two rubber quad rings. The quad rings provide a seal between the piston chamber and the tank and also keep the piston movement concentric within the cylinder. A rubber-coated steel ring located at the discharge pipe provides the soft surface required for the piston to seat and seal the cannon, again, at pressures as low as 40 lbf/in2 (2.7 bar). A piston-return spring can be located behind the piston in either type of valve to seat the piston immediately after discharge to prevent material from migrating back into the piston chamber.

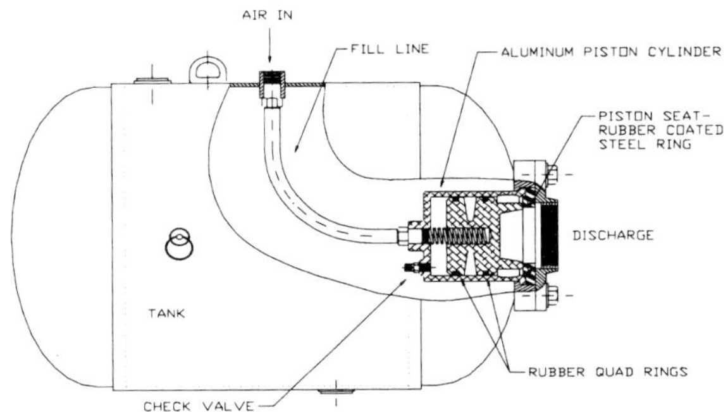

Figure 3. Internal Air Cannon

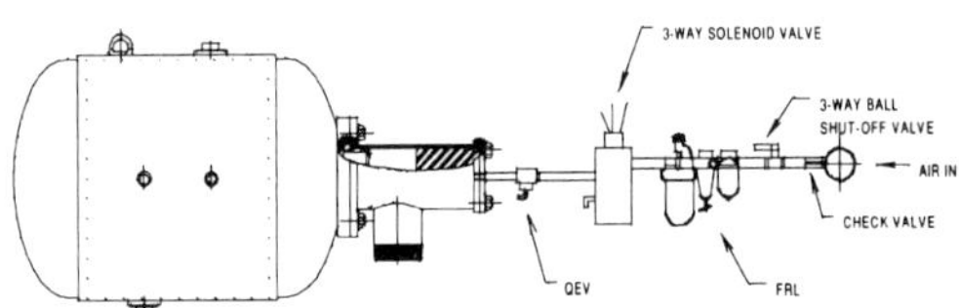

Figure 4. Air Cannon Components

1.2 The Air Cannon System

Figure 4 illustrates an air cannon system consisting of a check valve, shut-off valve, filter-regulator-lubricator (FRL), solenoid valve, quick-exhaust valve (QEV), control device, and cannon supply lines. The check valve and shut-off valve are placed downstream from the cannon to prevent accidental discharge and to allow maintenance of the system. The FRL provides the air preparation needed for long cannon operation and component life.

Either manual push buttons or programmable control devices control the cannon charging and discharging. The option of manual or automatic controls allows for the adaptation and customization required to meet changing material-flow characteristics and plant requirements.

Individual three-way, two-position, normally open solenoid valves are placed in the air line to operate the air cannon. If a spring is used behind the piston, solenoid manifolds can be utilized. The manifolds are housed inside a solenoid cabinet (Figure 5). Three or more solenoids can be placed in the cabinet.

The benefits of enclosing the solenoids into one cabinet are twofold. First, the cost of running conduit and wiring to each solenoid valve as required by other systems is reduced by four, since the wiring run is required only to each cabinet. Also, the air line components (i.e., FRL, shut-off valves, and check valves) are reduced by a factor of four since only one is required per cabinet. Installation costs of solenoid cabinets can be as much as 30% lower than individual solenoids.

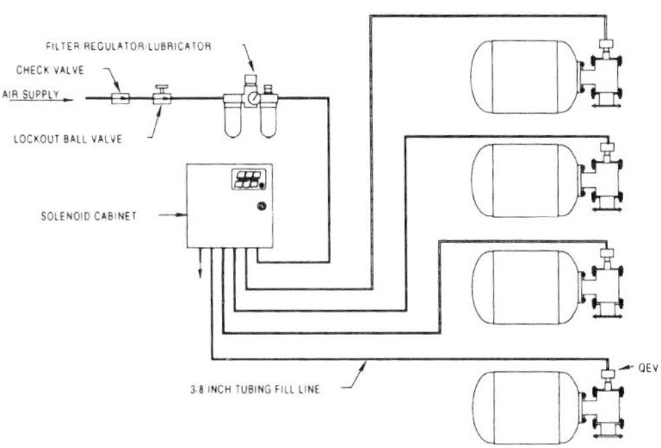

Figure 5. Multiple Air Canon Controls

Secondly, the solenoid valves are protected from harsh environments by the dust-tight, weather-tight enclosure that keeps dust and water from entering and causing failures. Optional enclosures can be specified to adapt to the harsher environment and to application needs.

1.3 Air Cannon Operation

External-valve and internal-valve cannon operations are similar and can be categorized into two phases.

1. Charging: The tank is filled with plant compressed air between 40 and 120 lbf/in2 (2.7 and 8.5 bar). A normally opened solenoid valve, which can be located away from the cannon, controls the filling. Air fed through the cannon-mounted QEV pushes the piston against the seat, sealing the tank. Air then flows into the tank through holes in the external-valves piston or through the check valve of the internal-valve design. When tank pressure equalizes with line pressure the airflow is static, and the cannon is ready for phase two.

2. Discharge: Activating the solenoid valve shuts off the air supply and empties the air line between the QEV and the solenoid, opening the QEV. The purpose of the QEV is to release

the pressure holding the piston. The piston is forced back instantly, releasing the energized air pressure stored in the tank. The subsequent air blast is directed through the discharge pipe into the material.

Following the discharge, the piston is returned (by the air line pressure or by spring pressure, if present) to prevent the backwash or migration of dust, material and gases from entering the tank. The full cycle time, from piston release to piston reseat, is less than half a second.

3. SIZING AND PLACEMENT OF AIR CANNONS

Selection of the proper type of air cannon discussed earlier is typically dependent on vessel temperature and type of material. Sizing of air cannons is dependent on many variables. These include:

- Size of silo or hopper
- Material density and moisture content
- Nature of flow problem (Bridging, Clinging, Rat holing, or Arching)
- Degree of slope of hopper wall
- Size and type of the discharge

Only when this information is known can the proper air cannon size be determined.

The majority of material flow problems occur close to the discharge. Placement of air cannons in this area is critical since this area must remain clear for the material above to flow. The vessel's discharge design is important because it inherently causes flow problems. For example in Figure 6, an open slot screw conveyor discharge can pull cohesive material in the direction of the screw travel. This compacts the material against the hopper wall, bridges it above the screw, and prevents even flow to the discharge. Another example is a feeder belt discharge. It, too, can pull the material to cause build up. However, if too large an air cannon is placed near the belt, the volume of air could lift up the belt sealing system, blasting material out of the stream of flow. This causes clean up and product loss issues. To prevent this situation, smaller air cannon sizes should be selected.

Once the discharge is assured to be open, the rest of the silo must be addressed to prevent flow problems. Additional levels of air cannons should be installed higher up the vessel walls to prevent bridging. The same variables mentioned above determine the coverage or reach of a particular air cannon size. In order to optimize an air cannon system, knowledge of the relationship of the application and cannon forces is necessary. If insufficient volume of air is introduced into the material, it may not move the material; if more volume is blasted into the vessel then needed to move the material, air is wasted and the cost of operation increases, along with the initial investment.

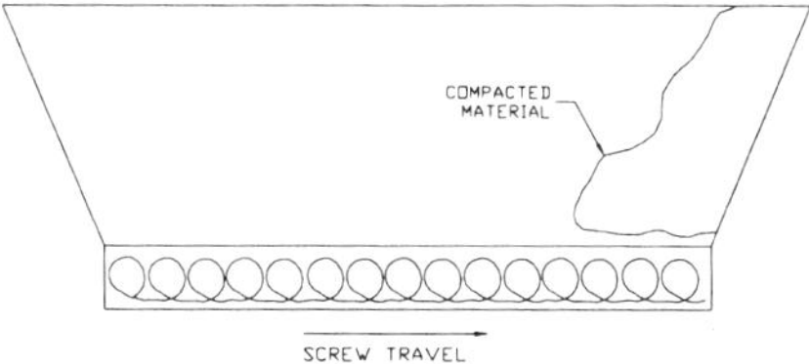

Figure 6. Screw Conveyor Problems

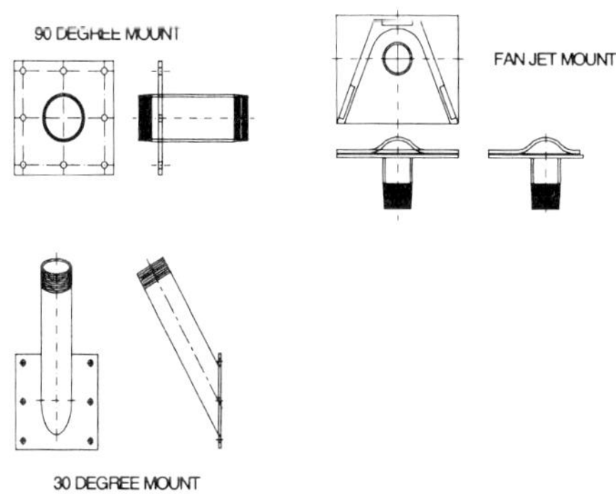

Figure 7. Air Cannon Mounts

Some processes experience a decrease in quality if the silo or hopper is not completely evacuated. In these cases, air cannons should be installed to cover the entire slope section, with the installation extending up the vertical section to dislodge any material that may be built up.

4. MOUNTING AIR CANNONS

Figure 7 shows the four basic types of cannon mounts are the 30-degree, 90-degree, fan-jet, and blow pipe.

The 30-degree mount plate is the most common and is used to provide a downward blast to eliminate bridging, arching, and rat holing.

The 90-degree mount plate provides a straight-in blast to break up bridging material or clean flat surfaces such as ledges.

The fan-jet mount directs the blast along storage-vessel walls to remove clinging material.

The blow pipe is used in high-temperature applications to remove the cannon from heat and prevent material from getting into the cannon. The blow pipe is typically welded to a fan jet to sweep adjacent walls of square or rectangular containers such as pre-heaters in cement plants.

4. CASE HISTORY: AIR CANNONS OVERCOME WINTER CHIP FLOW PROBLEMS

Production at a large paper mill in the Upper Midwest in the USA suffered when winter conditions hampered their wood chip handling process.

The plant moved purchased chips from outside stockpiles through the underground reclaim system's bins onto the conveyor that carried them to the digester. In the summer, the chip flow through these bins averaged 396 tons per day. But in the winter months, the cold weather would affect the chips, freezing them together or to the vessel walls, leading to slow flow and a production bottleneck. The plant's average daily chip handling dropped seven percent to 369 tons per day.

With a production value of $200 per ton of chips, this loss of 27 tons per day through inefficient chip movement cost the plant $5,400 (US) per day in lost capacity. The area experiences approximately 60 days of cold weather every year, resulting in a loss of over $300, 000.

In efforts to maintain the chip flow, the plant resorted to sending two or three workers to poke the material bottlenecks from the bottom with rods or an air lance. In some cases, employees were sent inside the bin or used live steam to free the material accumulations. All were inefficient procedures and, worse, potentially unsafe.

Seeking a way to improve chip flow and production efficiency, plant officials sought out an air cannon manufacturer to develop a solution. They developed an engineered solution of eight air cannons to reduce bridging and rat holing inside the hopper.

It was a goal of the mill to improve worker safety, without damaging the hopper. The air cannons offered a flow improvement, without requiring workers to climb the hopper to poke, hammer, air lance, or enter the vessel to dislodge the wood chips. Because the air cannon discharge acts on the wood chips rather than the hopper wall, there is no damage to the vessel.

The chip bin dimensions were 30 feet (9.1 m) wide, 50 feet (15.2 m) tall, and 30 feet (9.1 m) long and had a concrete sloped bottom. The eight air cannons were installed in three levels of three, two and three from bottom to top. The control system was tied to the operation of the bin paddle wheel feeders. The cannons fire on a timer only when the feeders are operating, pulling chips out of the bin onto a conveyor.

After the initial trial, the system's automatic sequence was set so each of the eight cannon fires once in a twelve-minute cycle. The cycle moves across the vessel, and then up the wall, firing one cannon every ninety seconds. Over the past winter, the air cannons improved flow so there was no appreciable difference between winter and summer capacity. The plant has noticed the flow improvement so much that they will be installing more air cannons on two additional chip storage bins.

In addition, the plant has now learned another benefit of the air cannon installation. Rather than just using the air cannons in the winter, the plant manually fires the system during the summer whenever hot, humid conditions cause the damp chips to stick together and create blockages. What was just seen as a winter solution has improved plant efficiency whenever chip flow leads to a production bottleneck?

With a $300 000 per year increase in production, the air cannon system was paid back quickly. Also, with minimal maintenance and repair costs, the system will increase the plant's production for years to come, giving them a safe, reliable solution to their material flow problem.

5. CONCLUSION

Air cannons provide powerful blasts of compressed air to porous and compactive materials to maximize the flow from storage or through processes. They solve flow problems on large storage vessels where the transmission of induced vibration fails and without expensive hopper redesign. The two basic types of air cannon--external-valve and internal-valve--provide similar operation and system designs. The use and application of air cannons and their mounting plates allow customization of the cannon system to solve material flow problems and meet the specific needs.

S604/003/98

External vibrators

I MATHER
Vibratechniques Limited, Brighton, UK

Synopsis

Vibration is the most commonly used discharge aid. Vibration can be very effective, but there are different types of vibrator with different effects. This paper seeks to clarify the different types, size selection and explains mounting procedures, other requirements and modifications, and looks at rotary, and linear, air and electric, and vibrations in respect of other discharge Aids.

1. INTRODUCTION

Using vibration to empty awkward hoppers and silos has been used since materials were first put into containers. When you tap the side of a container, the contents will spill out better. If you make this a rhythmic tapping, you can even induce a steady flow rate. Even back to beyond the ancient Greeks, grain, sand, aggregates etc., have been emptied from containers using a form of vibration.

Even in these modern times, the same method is regularly employed. Many a hopper is seen to suffer from what is commonly known as "hammer rash." This is where a person has hit the side of a hopper to induce flow. It therefore follows that a shock wave does help.

Some types of vibrator utilise this shock wave method, but more controlled and without undue damage to the hopper or silo. Other forms of vibrator use a more gentle approach and cause material flow by either making material close to the vibrating wall start an avalanche, or by reducing friction on the hopper wall inner face. In fact, all vibrators manage a little of all these methods, but have more effect in some ways than others.

2. TYPES OF VIBRATOR

In line with many machines, either their prime mover or the way in which they work can classify vibrators. Therefore, in common usage, we look at electric, air and hydraulically driven. We also look at rotary and linear. Of the types commonly available, most electric vibrators are rotary, and tend to be lower frequency. Pneumatic vibrators are commonly available in higher frequency rotary, and are the most common type of linear, (piston) vibrator. Hydraulic vibrators are commonly rotary, and commercially available ones tend to work at frequencies similar to their pneumatic counterparts.

2.1 Electric vibrators

Electric vibrators are normally a straightforward three-phase squirrel cage induction motor, like any other three phase brushless motor. Instead of the rotor shaft protruding out of the end of the motor to connect to, say a gearbox, the rotor shaft stays inside the casing, and eccentric weights are attached to them. When the motor spins, these out of balance weights produce a centrifugal force, which can be utilised as the useful vibration. Electric motors such as these spin at speeds dependent upon the frequency of the electric supply. It is quite common therefore to see 3000 r/min, 1500 r/min and so forth. (Using electronic inverters, we can get various speeds, but for hopper evacuation, such devices are not normally necessary.) Most hopper and silo electric vibrators work at 3000 r/min. This is because lower frequencies would, for a given force, give a much higher amplitude that could be detrimental to the hopper wall.

2.2 Pneumatic vibrators

Again, rotary vibrators are the most common pneumatic vibrator. Instead of eccentric weights attached to a shaft though they tend to be either a ball bearing spinning round a chase, a roller spinning round a chase or a turbine that has eccentric mass. (Holes in one semi circle, and heavy weight inserts in the other.) Because of the inefficiency of compressed air, they are normally the smaller type of vibrator, and need to spin at very high frequencies to achieve a working centrifugal force. Ball type and turbines can spin up to 40 000 r/min in the smaller vibrators.

Air is an excellent power source to achieve a linear piston vibration. A shuttle, not too dissimilar to a valve shuttle is, by the design of how the air ports around the casting, pushed from one end of the casting to the other rapidly. This reciprocating shuttle, moving very fast and then having to stop and change direction achieves a vibration. As you rely on the mass of the shuttle and the linear displacement, together with the rate of change of direction, the actual force, for a given physical size is not too great, but they are far more efficient users of air than their rotary counterpart.

One slightly different machine is a pneumatic hammer. In this, the shuttle is a large piston, pushed up a cylinder by air, held by either an internal spring or an air cushion. When the air is released, the piston accelerates down to a striking plate, or even the hopper wall when used with a seal, and sends a shock wave through the material.

2.3 Hydraulic and mechanical vibrators

Most hydraulic vibrators are similar in operation to pneumatic vibrators, but with the much higher torque available, can use lower rotating speeds, and heavier eccentric mass. In that case, they are a good compromise between electric and pneumatic. In practice, they are expensive to manufacture and use a rather high flow from a hydraulic pump. They are very seldomly used on hoppers or silos.

Pure mechanical vibrators fall into this category. They are often similar to electric vibrators, but powered by pulley wheels. This means that by reduction, higher frequencies are achieved. However, they are not used on hoppers, and with the advent of inverters, are no longer needed to achieve higher frequencies.

Almost all vibrators sold today throughout the world for hopper and silo discharge are either electric or air operated. It is not only a consideration of what power source is readily available, but also because of the differences in frequencies of the rotary vibrators, and the unique action of the linear piston vibrators, we must specify a vibrator on the needs of the application.

3. APPLICATION

Different granules of different material, dependent upon their size and density react better at different frequencies of vibration. They will resonate because they are "in tune" with a harmonic of their natural resonate frequency. Generally, the smaller the particle size, the higher the frequency of vibration to make enough particles resonate to cause migration. As particle size is not something we can control to the nth degree, a given frequency will normally affect enough material to have an effect.

3.1 The different effects

3.1.1 Shock wave

This does not discriminate between particle sizes, and is there to disturb the material by passing a wave of energy through it. It is produced most efficiently by a linear vibrator, (the moment of the force when a mass suddenly stops and goes the other way,) and is best utilised to break bridging within a hopper cone. It can, if powerful enough resettle material, meaning that rat holing and plugging of hopper cones can be affected by a shock wave.

A shock wave can be a single pulse or can reciprocate. (Impacting hammer vibrators work up to 3000 v/min.) In this case, some of the benefits of rotary vibration can be seen to work. Vibration is largely dependent upon force and frequency, not whether it is caused by centrifugal or linear means.

3.1.2 Migration

If we were to say that most vibrators on hoppers and silos are electric, then it is also true to say that migration is the most used method of hopper discharge by vibration. Most electric vibrators on hoppers are working at just under 3000 r/min. This relatively low fundamental frequency means that strong harmonics exist in the ranges where most materials are stored in granular sizes. In short, larger granules will resonate and move from the fundamental vibration, and smaller particles will resonate with the higher harmonics.

The net result is a movement, which once started, is aided by gravity to drop the material down. It gathers momentum and falls out of the bottom orifice of the hopper.

Migration of materials also leads to consolidation. Interlocking of larger grain sizes, and smaller grains filling the voids means that the material will compact. Therefore, warnings of never use a vibrator with the hopper bottom orifice shut are very pertinent warnings!

3.1.3 Lowering of friction in the hopper wall

Smaller pneumatic vibrators work at very high frequencies. Because of this, the amplitude of the vibration is low enough to be almost negligible. Also, the fundamental vibration, let alone any harmonics above that frequency will not affect anything larger than dust. You would be forgiven for thinking, therefore, that these vibrators are no good at emptying hoppers and silos. You would be wrong. Up to yet, we have been concentrating on the affect of vibration on the actual material. We have been ignoring the hopper wall.

Put simply, at a frequency of > 20 000 r/min, a rotary pneumatic vibrator will make a steel hopper wall very non-stick. As the amplitude is so low, it will not penetrate into the material too far, so is better as a preventative option rather than a cure. If a hopper wall is vibrating at such a high frequency, small material will not start to cake up on the wall.

Bearing in mind the precaution referred to in 3.1.2., regarding compaction, then as very small particles may migrate under high frequency vibration, then materials that are stored in very fine small particles size, such as cement, flour and pulverised fly ash may compact in a hopper subject to high frequency vibration.

3.2 Size selection

The three different main effects that vibration has means three different ways of specifying size. This is simplified by the fact that the three effects are achieved in the main by the three common types of vibrator.

- Electric vibrators for migration
- Pneumatic rotary vibrators for lowering of friction
- Linear pneumatic piston vibrators for shock wave.

3.2.1 Electric rotary vibrators

Generally speaking, we look at the mass of material in the hopper. (As the wall thickness and therefore mass of the hopper is designed to take the mass of material, and not much more, we can accommodate it in this simple equation.)

Between 10% and 15% of this mass, expressed as weight is the equivalent force required from a 3000 r/min vibrator.

Vibrators are rated in force, normally Newtons (N) or, incorrectly but conveniently Kilograms of force, (KgF.) As there are 9.81 N in 1 Kg of mass wherever earth's gravity is felt, we can talk about Kilograms of force without fear of contradiction.

This is useful for whoever is specifying a vibrator, or series of vibrators. For instance, a VIBTEC MVSI3/800 type vibrator means 800 KgF at 3000 r/min. A Wacker AR15 means 1500 KgF, and so forth.

If you have a hopper that holds 1 tonne of material, then the force required can be expressed as follows;

Working on the 10% rule.

Material weight = 1000 Kg. 10% of this is 100 Kg.

Force required is 100 KgF.

(Working on the 15% rule, a vibrator achieving 150 KgF is required.)

The rule you work to depends upon how awkward the material is to discharge. Other factors could be to fit in with a given size of vibrator. Using the VIBTEC range, the hopper above would be best suited to an MVSI3/200, (200 KgF.) As the cost of a 300 KgF vibrator, (MVSI3/300,) is negligibly more expensive, then this is the machine best suited. After all, you can adjust the weights to give a lower force output. This would serve to help with the lifespan of the vibrator. To give 100 KgF, you will be running the vibrator at only 33% of its capability.

3.2.2 Pneumatic rotary vibrators

With pneumatic rotary vibrators, you are working between a lowering of friction and migration. The smaller the vibrator, normally the higher the frequency and therefore the more of friction reduction.

The similar rules apply, as per electric vibrators, only they are far better suited to smaller grain sized materials. Large grain sizes such as >10mm limestone etc., will see little effect from high frequency vibrators. However, a reduction in the friction of the hopper wall will help to stop blockage in the first place. If this is the case, then a much smaller vibrator is required, and we can ignore the 10% to 15% rule, and use a smaller vibrator.

Pneumatic vibrators can have their force and frequency adjusted by regulating the pressure of the incoming air supply. As the relationship between centrifugal force and frequency is a square law, halving the frequency will result in only a quarter of the force. This means that size selection is not too important. A wide variation of forces and frequencies are available from any given vibrator.

3.2.3 Pneumatic piston vibrators

Although there may be ways of specifying the size by more scientific means, any experimentation to find a pattern tends to show that the crude normal method works. The size of an impacting piston is selected by the thickness of the hopper wall.

We actually size the vibrator by the diameter of the piston. The pressure of the resultant shock wave depends on the force of the piston hitting the striking plate, divided by the cross sectional area of the piston. Judging by the diameter is an effective way. (The displacement of the piston as it travels and the rate of acceleration all goes to give the force of it hitting the plate.)

A reciprocating piston tends to be a lower pressure shock wave, but the frequency of the waves mean that a migration inducing vibration is also assisting.

3.3 Mounting of vibrators

3.3.1 Rotary vibrators

Rotary vibrators are mounted onto bearer channels, stretching the length of the cone of the hopper. These bearer channels must be stitch welded, preferably using low hydrogen welding rods or a MIG welder to give a low tensile malleable weld. Standard "C" channel is adequate, mounted toes down. The vibrator is then mounted on a flat plate that is firmly welded to the channel. If a very large vibrator is required, two channels are most suited mounted in parallel, back to back, i.e., toes out. The gap between them being determined by the pitch of the hole centres of the vibrator, so that of the four bolt holes, two are drilled through each channel. Again, a flat mounting plate is required between channels and vibrator.

The vibrator is found to be best mounted approximately one third of the way up the cone from base. The centre line of the rotor, (or turbine, roller etc.,) being transverse to the length of the channel.

With electric vibrators, it is most important to check the running current of the vibrator when first installed. If it exceeds the nameplate current, reduce the force by adjusting the eccentric weights until the current is within specification.

Pneumatic vibrators will exhibit an excessive lateral motion if the force is too high. This leads to high wear and short life. If lateral motion can be seen with the naked eye, then lower the applied air pressure.

3.3.2 Piston vibrators

These need to be mounted on a welded pad or on a short length of channel, mounted toes down on the hopper wall. The positioning is best decided by the application. If there is a bridge to break, then mount it at the height the bridge occurs. If you are using it to clear a hopper wall, then again one third of the way up the hopper wall is found to be most beneficial.

4 CONCLUSION

This paper has set out to destroy a few myths about vibrator selection and mounting. It has also tried to put forward a case for why materials react in the way they do when subjected to vibration.

If the practical advice contained within this paper were applied, the familiar stories of burned out electric vibrators, badly worn pneumatic vibrators and holes in hoppers where a vibrator was mounted directly to the skin would all be stories of the past.

S604/004/98

'Soliflo' vibratory hopper discharger

G ELLIS
Solitec Handling Limited, Gloucester, UK

1.0 INTRODUCTION

Most Industries have a requirement to store particulate solids in bulk. This material then has to be discharged at a consistent, controlled rate to process.

Because of the huge variance in the flowability of materials, each system has to be carefully engineered to ensure trouble free discharge.

1.1 Material Flowability

Many parameters affect the material flowability.
Density.
Particle Size and Distribution.
Particle Shape.
Moisture Content.
Fat/Oil Content.
Temperature.
Sensitive to Aeration.
Sensitive to Compaction.
Sensitive to Time Consolidation.

1.2 Other Application Factors Affecting Flowability

Volume of Material Stored.
Storage Hopper Size and Shape.
Inside or outside of building.
Length of storage time.
Method of hopper fill.
Type of take away equipment.
Discharge Rate.

1.3

It can be seen that an apparently simple requirement can lead to site problems if not properly engineered.

2.0 TYPES OF FLOW

Basically, there are Two Types of Flow from a Storage Hopper.

(a) Mass Flow.
(b) Core Flow.

See Illustration 1.

2.1 Mass Flow

Mass flow occurs when the material moves down the vertical barrel of the hopper as a solid mass. There is little relative movement between the particles, and the top surface retains its original configuration, i.e. the material slides over the walls of the hopper in preference to sliding over itself at a place remote from the walls. This gives a first in, first out principle.

2.2 Core Flow

Core flow occurs when the material fails, and slides over itself.

The top surface indents and flow takes place by the upper layers funnelling into the centre, moving down to the outlet in a moving core. First in is not necessarily First out.

Depending upon material properties, a proportion can remain in the hopper and segregation can occur.

2.3

It can be seen from the Illustration that mass flow hoppers give a superior flow pattern, and is essential when discharging into a metering device.

3.0 DISCHARGE METHODS

There are a number of discharging methods which can be utilised to promote material flow from storage.

3.1
Hoppers designed to give gravity flow without the use of any additional discharge aid.

3.2 Aeration

3.3 Mechanical Dischargers

3.4 Vibratory Dischargers

3.5
Each discharge method has advantages and disadvantages.

It is important at the initial design stage to select the correct discharge method for the particular Application being considered.

4.0 DISCHARGE PROBLEMS

See Illustration 2.

The subject of this paper is the "Soliflo" Vibratory Hopper Discharger.

Before considering the design and operation of the Soliflo, it is expedient to consider discharge problems. It can then be shown how the "Soliflo" assists in overcoming these problems.

4.1 Bridging
This is caused by the increase of the shear stress of the material, when it is being funnelled into a progressively narrowing outlet. The harder the material is compacted into a confined space, the greater the friction between the particles until the material becomes so strong that it will not shear, and bridging takes place.

4.2 Rat-Holing
Because material is being discharged from the outlet in a continuous stream, the shear strength of the material above the outlet is not allowed to increase, and this column of material continues to flow freely. The material being supported from the sloping walls of the hopper, is compacted between the hopper walls and the free flowing column. The material shear strength in this area increases, and flow will not take place in this area.

4.3 Segregation
This generally occurs during the hopper fill. It is a natural phenomenon that when materials form a pyramid shape, such as occurs in the filling process, the larger particles roll to the outside, and the small particles stay in the centre.

4.4 Degradation
This can be caused in Three different ways -

(A) The breakage of product by the filling process.
(B) The passage of material against itself, such as occurs in a core flow condition.
(C) Mechanical work performed on the product by the discharging device.

4.5 Flushing
Certain micron particle size materials readily aerate, and in this condition, take on the temporary characteristics of a fluid. The more thorough the aeration, the more fluid to the material. In this state, accurate metering is difficult to achieve, and control can easily be lost.

5.0

Having examined the Two Types of flow pattern, and identified the various problems associated with discharging materials from storage, the next step is to overcome these problems.

One Method is by use of the "Soliflo" Vibratory Discharger.

6.0 "SOLIFLO" VIBRATORY DISCHARGER

6.1 Introduction
The "Soliflo" is a pre-assembled, self contained machine, which when fitted to the bottom of a silo/hopper gives a controlled, variable discharge rate of material from storage, with positive shut off.

Vibration is a standard feature of the "Soliflo". The mode of vibration is in the vertical plane. This means that whilst round and square "Soliflos" are standard, any shape can be considered to suit an existing silo/hopper.

"Soliflos" are provided in both Carbon Steel and various grades of Stainless Steel, with special coatings, i.e. P.T.F.E., Epoxy etc. to order.

6.2 Description
The "Soliflo" is a conical shaped machine with the body split into Three Parts.

The transition cone (5), the main body (4) and the outlet cone (9).

The transition cone (5) has a connection flange (8) welded at the top, which is bolted to the outlet flange of the hopper/silo.

The outlet cone (9) is manufactured with an outlet configuration to match the piece of equipment taking material away from the "Soliflo".

The main body (9) has a support system for the operating mechanism. This is welded to the main body by a cruciform framework.

The base of the pneumatic actuator (1) is bolted to the top plate of the support system.

The top of the pneumatic actuator (1) is bolted to the vibrating top cone (3).

Sandwiched between the pneumatic actuator and vibrating top cone is an isolation gasket (6).

The vibrating top cone (3) contains an air operated vibrator (2) at its apex, and a rubber seal (10) around its bottom edge.

When the vibrating top cone (3) is in its rest position, the rubber seal (10) forms a dust-tight connection with the transition cone (5).

Attached to the lower supporting flange of the vibrating top cone (3) are height adjusting rods, each fitted with an adjustable collar (7) at its lower end.

The "Soliflo" is operated pneumatically, and is therefore intrinsically safe.

All pneumatic connections are exhausted externally to ensure the operating air does not become in contact with the stored material.

6.3 Installation

All "Soliflos" are designed to be bolted directly to a prepared flange on the hopper/silo outlet. The only power supply required is an air line with a maximum pressure of 7 bar g. The air supply must be clean and dry.

A control panel is supplied with the "Soliflo". Contained within the panel is a regulator to enable air pressure to the pneumatic actuator (1) and the pneumatic vibrator (2) to be adjusted. The panel also contains a lubricator for the air supply to the pneumatic vibrator (2).

The pneumatic actuator (1) requires clean, dry air, and the pneumatic vibrator (2) requires clean filtered and lubricated air.

The control panel must be mounted within 3 Metres of the "Soliflo".

6.4 Operation

Once the "Soliflo" is installed correctly, the only control required is on/off air to the vibrator and actuator.

When air is injected into the pneumatic actuator, it inflates and lifts the vibrating top cone vertically, until the adjustable collars connect with the underside of the pneumatic actuator bottom support plate. This creates an annulus through which the stored material can flow.

By adjusting the position of the collars on the height adjusting rods, this annulus can be varied to control the material discharge rate. The wider the annulus, the greater the discharge rate. This lifting action also has the effect of disturbing the material immediately above the vibrating top cone and assists in breaking any bridge which may have formed.

Air can now be injected into the pneumatic vibrator allowing vibration of the top cone.

These vibrations are transmitted from the cone into the material.

Under controlled laboratory conditions, it has been found that under vibration, the shear strength of a powder bed, decreases with the increase in packing density. In other words, the vibration imparted reduces :-

(A) Friction between the top cone and material.
(B) Friction between individual particles of materials.
(C) Friction between the material and wall of the storage hopper/silo.

This decrease in material shear strength promotes material flow.

At the "Soliflo" outlet, the material is in a flowing state and therefore of low mechanical strength. It is thus able to pass through the relatively small orifice at the outlet.

A combination of annulus and vibration will promote material flow from storage.

When discharge is no longer required, the air supply to the vibrator and actuator is closed. Air within the actuator is vented through quick exhaust valves, lowering the vibrating top cone into its sealed position. By careful selection of "Soliflo" size, annulus size and vibration, it is possible to achieve mass flow of material from storage.

6.5 Service

If during operation a fault occurs within the "Soliflo", the vibrating top cone (3) can be lowered to form a seal with the transition cone.

Unbolting the flange connecting the main body (4) to the transition cone (5), allows access to the operating mechanism.

The total assembly can now be removed for servicing, leaving the vibrating top cone (3) in position.

This means servicing can take place on the "Soliflo" without emptying the hopper/silo.

7.0 OVERCOMING DISCHARGE PROBLEMS

The "Soliflo" will assist in overcoming the Five storage problems previously described in Section 4.

7.1 Bridging

This problem is solved in Two ways.

(A) By passing vibrations through the material, the material shear strength is reduced. This allows flow to take place where normally compaction and therefore bridging would occur.

(B) Cutting away the hopper of the storage vessel, near to the point where bridging would normally take place. Because of reduction in material strength, it enables a large diameter "Soliflo" to be utilised with a small diameter outlet.

7.2 Rat-Holing

To Rat-Hole, the material has to flow down the centre of the hopper/silo, and through the outlet. The vibrating cone of the "Soliflo" prevents this from taking place.

Additionally, by reducing the material strength across the stored bed, flow is generated at the hopper/silo wall as well as the centre core.

7.3 Segregation

Because flow is generated across the material bed, the velocity of the material at the hopper/silo wall, is similar to the material velocity of the centre core. This tends to give a re-mixing action at the annulus.

7.4 Degradation

As described in 7.2 and 7.3, the material tends to drop en-masse. i.e. mass flow. There is little movement of material against itself, and prevents one cause of degradation.

Additionally, minimal mechanical work is performed on the material by the "Soliflo" action.

7.5 Flushing

The vibrating action of the "Soliflo" has the effect of de-aerating the material, eliminating the cause of flushing.

7.6
It can be seen that this one machine assists in eliminating all Five problems associated with the discharge of material from storage.

8.0 "SOLIFLO" SIZE

The "Soliflo" is a simple machine, but the application for each specific installation is complex and has to be examined most carefully to ensure satisfactory operation.

8.1
Generally the "SOLIFLO" diameter is a ratio of the diameter of the barrel section of the storage silo/hopper. It can vary between one quarter of the silo/hopper barrel diameter to the same diameter of the barrel (Full live bottom).

The more free flowing the material, the smaller the "SOLIFLO".

Normally free flowing materials are those which will readily transmit vibrations.

It is important that the vibration generated by the "Soliflo" vibrating top cone is transmitted via the stored material to the silo/hopper walls. This is necessary to ensure mass flow.

There are a number of factors to be considered to arrive at an optimum "Soliflo" size.

8.2 Material Flowability
Based on material flowability, an estimate has to be made at the approximate area in the storage silo/hopper, where the stored material is likely to bridge.

This area is virtually impossible to calculate, but with experience, an approximate bridging point can be predicted for materials which are commonly handled.

Materials which are unknown can be compared with a common known material, either by the examination of a sighting sample or by trials with a test "Soliflo" machine.

Once the bridging point has been predicted, this sets an initial size for the "Soliflo" required.

For example if the predicted material bridging point is 1000 mm diameter, the next standard size "Soliflo" above 1000 mm is selected.

8.3 Material Head Load
To allow material to flow, the "Soliflo" vibrating top cone must be raised to create an annulus. The pneumatic actuator selected must be powerful enough to overcome the head load of material upon it.

There are a number of accepted formulae available to enable the head load to be calculated.

In addition, an allowance must be made for power required to shear the material as the vibrating top cone rises. Once again experience is required to estimate this additional power requirement. If for example the material stored is free flowing, the additional power required is small. If the material is cohesive greater power is required. If the material is in a prill, fibre or pellet form, all of which interlock, it may be necessary to select a pneumatic actuator capable of lifting the total material weight in the silo/hopper.

Once the power required is known, it can be compared with the standard pneumatic actuator fitted to the size of "Soliflo" previously selected. If insufficient power is available in the size of machine selected, it may be necessary to chose a larger size "Soliflo".

8.4 Discharge Rate

From experience, the minimum and maximum discharge rates form a given size of "Soliflo" is known. The throughput required for a specific application can be checked against the known rate of the "Soliflo" selected.

8.5 Outlet

The size of outlet is normally specified to suit the piece of equipment the "Soliflo" is discharging into.

This outlet diameter must be checked against the material discharge rate required, to ensure it is large enough to pass this rate.

8.6

To summarise, the machine design is simple, but the application for a given specification is complex.

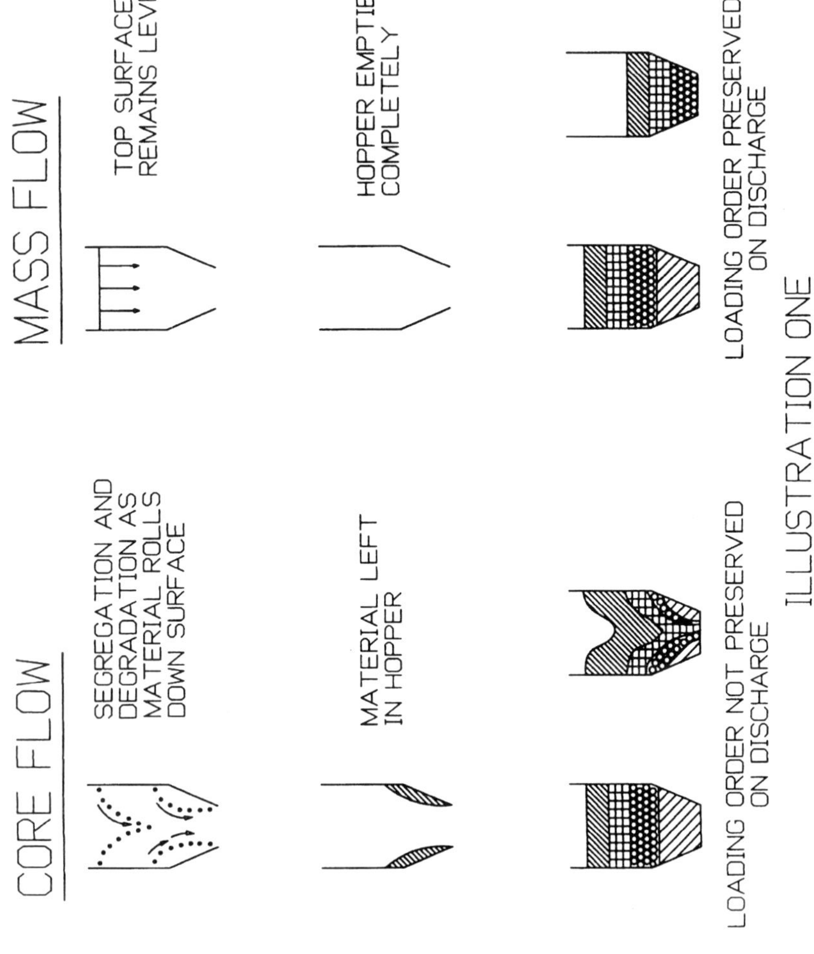

ILLUSTRATION ONE

RATHOLING

SEGREGATION

DEGRADATION

BRIDGING

FLUSHING

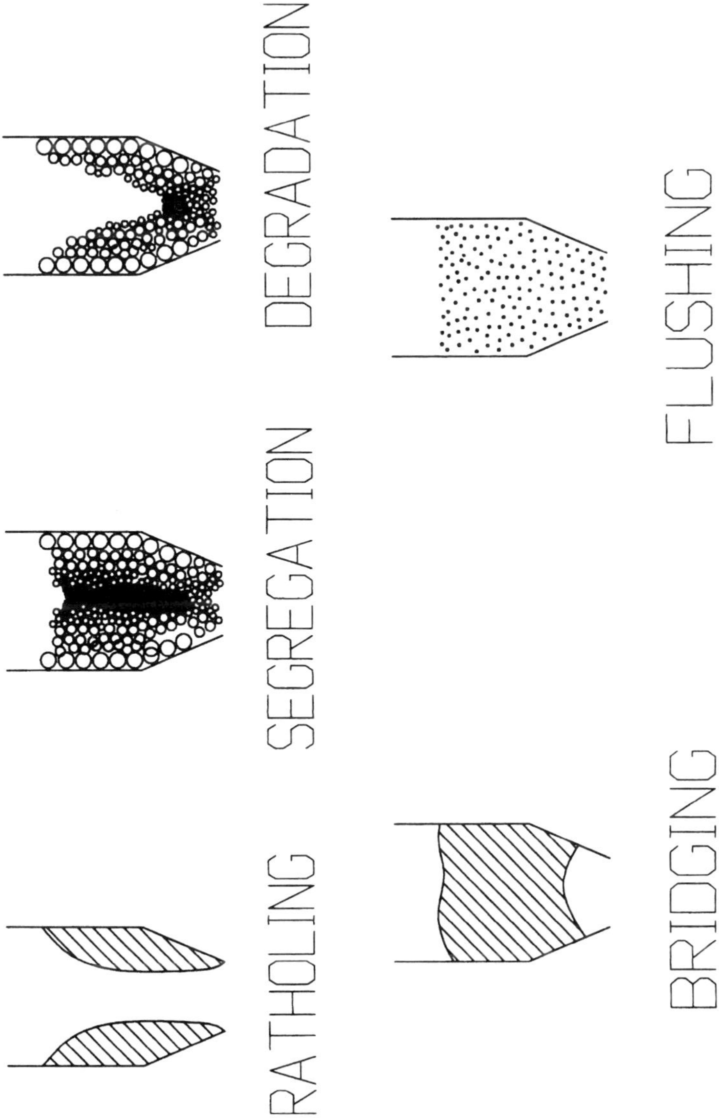

ILLUSTRATION TWO

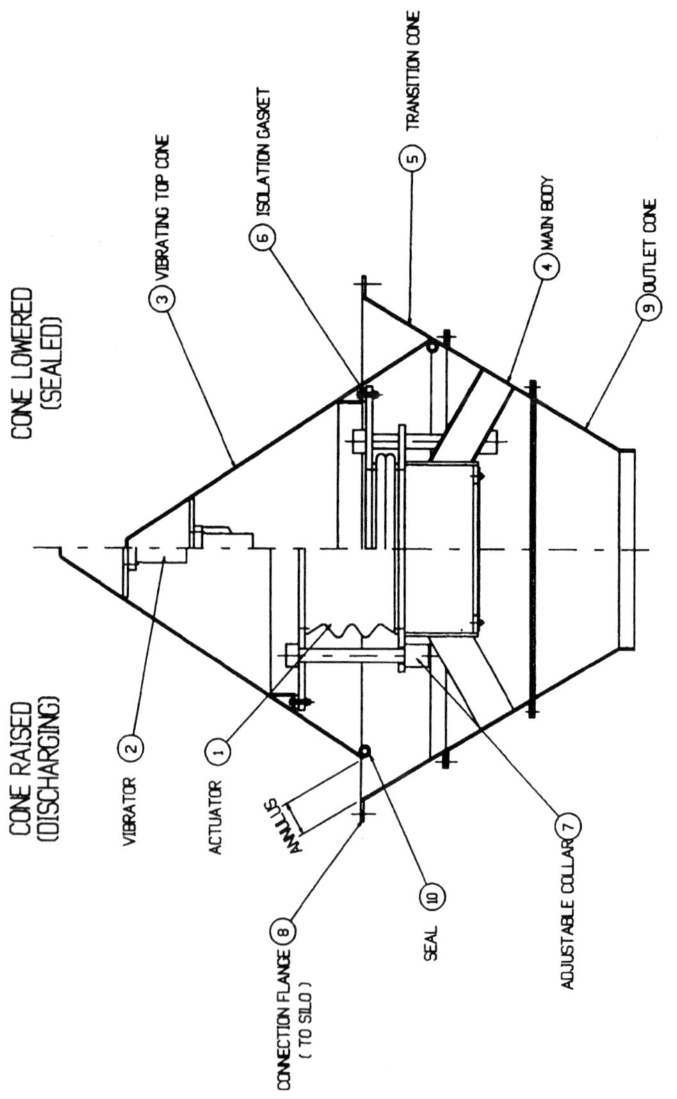

GENERAL ASSEMBLY
SOLIFLO VIBRATORY DISCHARGER

S604/005/98

Saxlund International flat bottom silo systems

F BOYLES
Saxlund International Limited, Southampton, UK

SYNOPSIS

When it is required to store bulk solid materials that do not readily flow, the Saxlund International approach is to store on a flat floor and design a reciprocating type of scraper machine that will 'dig' the material from the store. This store can be a silo, bunker, stockpile etc. We therefore have 'First in – First out'. This digging machine is a hydraulically operated frame or frames manufactured from special profiles. The stored material is thereby loosened and delivered into a screw or similar that will meter into the process etc.

1. INTRODUCTION

Mr Saxlund is a Norwegian engineer who has lived now for many years in Germany. He realised some 50 years ago that there was an ever growing demand for silo and bunker systems for timber residue materials for use in boilers as well as the then embryonic wood particle board industries. These wood materials, wet hacked chips, wet and dry flakes, timber residues, shredded bark and even sawdust and sanding dust will not readily flow from a conventional conical bottom silo. It was therefore necessary to start with a flat bottom silo and design a discharge machine that would reliably dig the stored material out. The first approach was with rotating machines of various types, the most famous being the Hydraulic Rotor which is still used today throughout the world by some of our customers in the particle board industry. This machine has hydraulically operated arms mounted on a rotating hub in the centre of the flat silo floor. When required these arms are driven out into the stored material and dig it out and at the same time they push it down slots in the floor. Beneath these slots is a progressive pitch screw conveyor that is maintained full of loose material. In this way it is able to give an accurate metering function of the stored material into the particular process. Some other types of circular discharge machine have be produced and tried over the years and this includes our hydraulically driven screw rotor machine. It has been concluded that although this design of machine can be made to work and is cheaper to produce, it has a number of inherent problems and is not as reliable in process applications as our reciprocating discharge machines which will be described later.

Some years later the same timber industries wanted bunkers with a rectangular format where this Hydraulic Rotor discharger would not be ideal. A new type of machine would have to be invented. The final solution was perhaps the most simple thing you could ever imagine. So simple in fact that even today when we have many hundreds in operation world-wide, people see it when it just installed and question whether it will work at all. This is our Push Floor machine.

THE 'PUSH FLOOR' MACHINE

The 'PUSH FLOOR' is based on a simple wedge shaped steel bar that today we purchase directly from the steel rolling mills so that it is hard faced, durable and relatively inexpensive. This is cut into lengths of some 1.4 meters and welded between solid rails to form ladders. These ladders which we call 'Pusher Frames' are then positioned on the bunker flat floor and driven in reciprocating motion by a simple hydraulic system.

When the stored material piled on top of these 'Pusher Frames' it is progressively moved to one end. This can be either end of the bunker or indeed into the centre where again we have our progressive pitched metering screw conveyor(s) positioned underneath, generally through a slot(s) in the floor. Today we have as many as 30 hybrid versions of this simple discharge machine operating with materials as diverse as de-watered sludge cake, gypsum rock, nylon flock, car tyres as well as many, many others. With some of these materials of course, it is not possible to use a screw conveyor as the discharge machine from the Push Floor. Sometimes we use a belt, sometimes a roller conveyor and sometimes a chain etc.

The Push Floor machine is an extremely useful device and it can be designed to do a variety of different jobs. Not only for silos and bunkers but also for 'drive in' reception with no expensive pit required and even as a stockpile reclaimer. They always give a 'first in, first out' material flow pattern, as indeed do all of our machines. As a general rule, the Push Floor machine, when used for bunker discharge, is suitable for storage capacities of up to 1000m3.

THE 'SLIDING FRAME' MACHINE

The next interesting development was to think of how to use this simple new idea in a circular silo to give all of the advantages of the simple but effective concept of the Push Floor to a round silo discharge machine. This was achieved by rolling the pusher bar material with its pointed edge on the outside and welding it into a frame so that when again driven hydraulically in reciprocating motion while sitting on the silo flat floor, it would dig and drag the stored material to the centre. In the centre it was then possible to make a wide slot right across the silo diameter. Remember there is now no rotor in the centre to get in the way. Again our famous progressive pitch screw conveyor was positioned underneath which was kept full of loosened material by the reciprocating action of the frame. Indeed, we can have more than one slot with more than one discharge screw if this is a requirement. Each discharge screw can perform a different duty if necessary.

This machine became known for obvious reasons as the Sliding Frame. The main advantage of the Sliding Frame apart from its simplicity and consequent modest price is its ease of maintenance and very low wear and associated costs. All maintenance parts are outside the silo to allow ease of access. Today the working parts of each frame, that is to say the sections with the pointed edges are produced for us under an exclusive agreement by a German steel supplier. They use a forged rolling process that gives a tough, long lasting and strong frame material. The fabrication of each Sliding Frame is governed by strict procedures and it is inspected at every stage during manufacture in our works. The design of each frame is specific to the particular material being handled, the silo diameter and the discharge rate. It is a very simple looking machine that some competitors now try to copy but of course we have more than twenty five years experience in its design, construction and its variations. The point where the driving rod passes through the silo wall must be sealed and this again is achieved using different proven seal types according to the material being handled. Sometimes with larger silo diameters and heavier bulk solids, we drive the frame using two hydraulic cylinders, one at each side of the silo. Also today we have a new patent for a frame that is made in two pieces which will give some advantages under certain conditions.

SUMMARY.

To summarise, we consider ourselves to be working at the leading edge of bulk solids handling technology for non free flowing and difficult materials. Today we construct many complete systems where we include sludge cake pumping and conveying machines. We also specialise in heat and power generation systems that use non fossil fuels. Here we supply either part or complete plants which often includes the complete fuel handling and ash handling systems. We see more and more difficult materials from time to time that our client's ask us to handle and each case is carefully considered by our research department before proceeding. So far we have found very few bulk solid materials that we are unable to handle with our simple pusher technology.

The Saxlund International 'Push Floor'

Discharge Machine for Rectangular Bunkers

The Push Floor is an 'original' discharger design by Saxlund International for square and rectangular bunkers. It is designed to function with non free flowing and difficult to handle bulk solid materials. The flat bottom floor concept gives many advantages and the Push Floor can be used in a number of different configurations to suit the client's particular requirement. Reliable discharge and accurate metering can be achieved on demand.

Reliable and versatile

The Push Floor discharge machine can be used in a number of different configurations. It can be used in the bottom of a square or rectangular silo or bunker. It can be used to push the material to one end of the bunker or to pull it to the other end. It can also push and pull the material to the centre of the bunker where there can be either one or several screws, solids handling pumps etc. It is then performing as a rectangular version of the Sliding Frame machine. The Push Floor can also be used as a 'drive in, tip and leave' machine or indeed as a stockpile reclaimer.

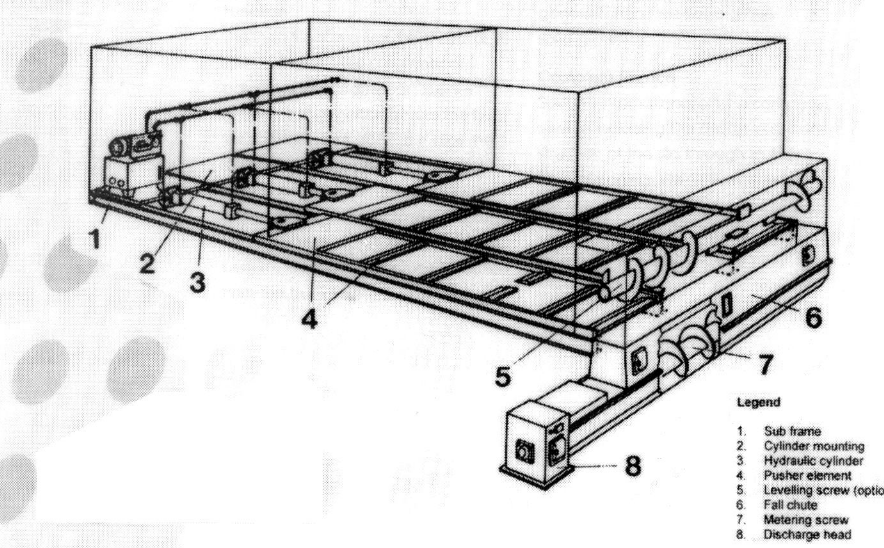

The Saxlund International 'Sliding Frame'

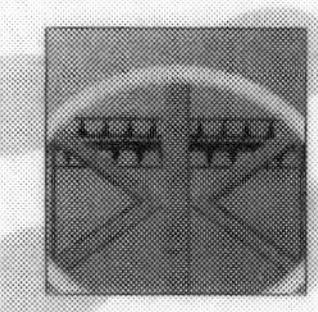

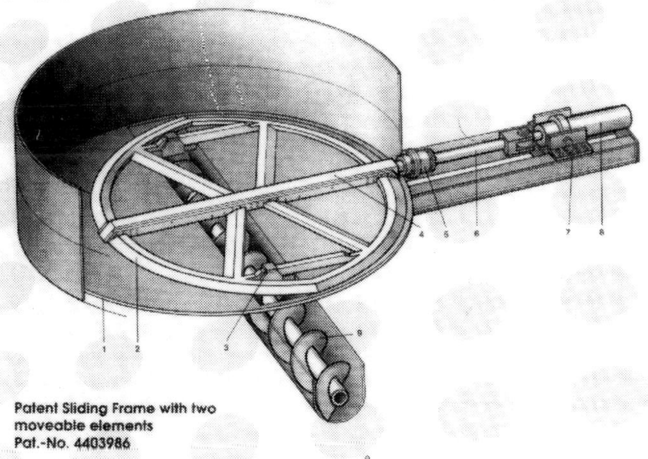

Patent Sliding Frame with two moveable elements
Pat.-No. 4403986

Legend
1 Silo bottom
2 Sliding frame
3 Frame guides
4 Central spine
5 Stuffing box
6 Connecting rod
7 Cylinder mounting
8 Hydraulic cylinder
9 Metering screw

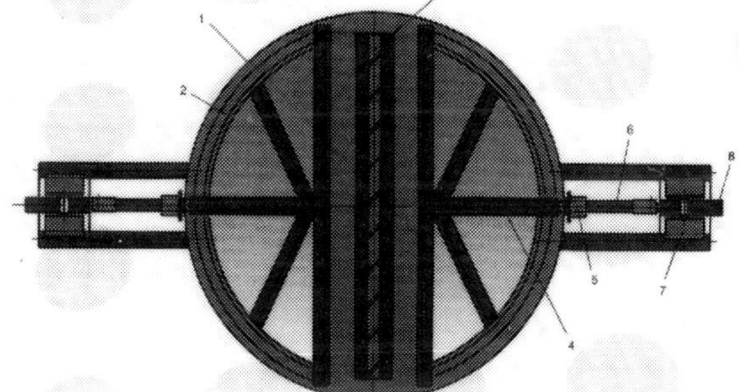

The Sliding Frame is an 'original' discharger design by Saxlund for flat bottom circular silos. It is designed to function with non free flowing and difficult materials such as dewatered sludge cakes. The flat silo floor concept gives many advantages such as the maximum possible sized discharge openings. The Sliding Frame discharger creates 'mass flow' within the silo even with these difficult materials. The client can be sure of achieving accurate discharge and metering of the stored material on demand no matter what the application.

Advantages
- **Completely enclosed - no odour**
- **Effective and simple operation**
- **Low power consumption**
- **Low maintenance costs**
- **Accurate metering**
- **Flat bottom silos are economic to manufacture and maximise storage capacity**
- **Complete systems from one experienced supplier**

S604/006/98

The Redler circular bin discharger

K WALKER
Redler Limited, Stroud, UK

Synopsis

The subject matter is a mechanical device which will break product 'arching' and provide a guaranteed continuous flow of product from a cylindrical storage bin.

While this discharger may be loosely linked with other devices as a flow aid it must be noted that it is a machine designed specifically for discharging poor flow materials i.e. those that would arch in a conical bin bottom and is normally included at the initial design stage of a handling and storage system. Retrofitting may be more involved than with other flow aids.

The **REDLER** Circular Bin Discharger has been produced by **REDLER LTD** for some 45 years and has had considerable success during this time. Two sizes are currently produced these being 800mm and 1000mm diameter and the design is based on fitting to circular bins having a conical bin bottom with a 50° flank angle. The bins may be up to 12m diameter and have a barrel aspect ratio of 2-3 (height/diameter). Within the two sizes are three types designated single stage, two stage and three stage. These are selected according to the required duty and the material to be handled.

The basic discharger is the single stage machine. The output from this discharger is dependant on the nature of the material and it's flow/breakaway characteristics. The bin is discharged, but not at a controlled rate. Should a controlled rate of discharge be required then a separate metering unit is necessary at the outlet of the discharger. However, high discharge rates of up to 200 t/hr have been achieved with single stage machines.

Many applications require more modest, controlled outputs. A two stage machine covers this requirement by adding a volumetric metering stage so that material discharged by the first stage is controlled by what is effectively a circular **REDLER** conveyor. Typically the discharge rates here are 5-50 tonne/hour.

Occasionally a three stage machine which has two metering stages is necessary to provide a greater degree of control with materials prone to fluidise.

The Circular Bin Discharger can be described as having two main elements; that which is internal to the bin being discharged, and that which is external to it.

The internal element is an archbreaker arm, which lies up the full slope length of the cone of the bin. It is connected to the rotating first stage by a heavy duty universal joint and is thus able to rotate on its own axis inside the bin. As a consequence the arm is not forced into the material, but is reliant on its own speed and cutting action to break down the arch which tends to form in the cone of a circular bin. The arm comprises a series of flights attached to a central bar whose section is dependent upon the nature and duty of the machine. The top end of the bar is supported on a 'spider' which runs around the upper circumference of the cone at the bin's 'hip'. It is normal to provide a wear strip at this point.

It is the arm which ensures that feed material is maintained to the external element of the Circular Bin Discharger at the base of the cone. This contains a rotating centre shaft which drives a centre housing to which flights are attached. These flights convey material to the discharger outlet.

When a regulated rate is required two stage machines are used. The additional stage beneath the first stage meters the material volumetrically in an annular space and no further measuring or control equipment is needed. As mentioned previously, finer materials which tend to fluidise or aerate may require the three-stage machine for effective control and metering.

The single or two-stage machines are normally fabricated from plate and rolled sections but the three-stage discharger incorporates machined castings to maintain close clearances. Normal drive units consist of geared motors or foot mounted motors coupled to a separate gear unit with an encased chain drive incorporating a shear pin device.

When variable discharge rates are required, to meet process plant requirements, variable speed drives can be incorporated with control from a remote operating position.

It is important to recognise the flow pattern of material being discharged from a bin by a **REDLER** Circular Bin Discharger. It is **not** a first in - first out pattern but core flow that varies according to the material characteristics.

If there is a feed into the bin at the same time as material is being discharged then a preferential flow channel is likely to exist resulting in the discharge of fresh material. This is an important feature to note if the material is likely to 'age' or go off. It is thus important with such materials to allow the bin to empty at regular intervals. Also as the action of the archbreaker arm is to cause undercut and collapse situations the bins need to be designed to allow for these variable loadings.

The **REDLER** Circular Bin Discharger is a mechanical device which has proven itself over the years to be effective over a wide range of materials with greatly differing flow properties, such as: - wood chips, leather shavings, lime, cement, spent hops, china clay, animal feeds, polystyrene chips, wheat germ, peat, soda ash, coal, boiler ash, sugar and phosphate rock to mention a few. Its design is uncomplicated and it is robust and reliable, requiring the minimum of maintenance. An additional advantage is that it generates very little noise apart from that of its motor and gearbox. With some material which will neither aerate nor transmit vibrations, it is the only suitable type of discharger.

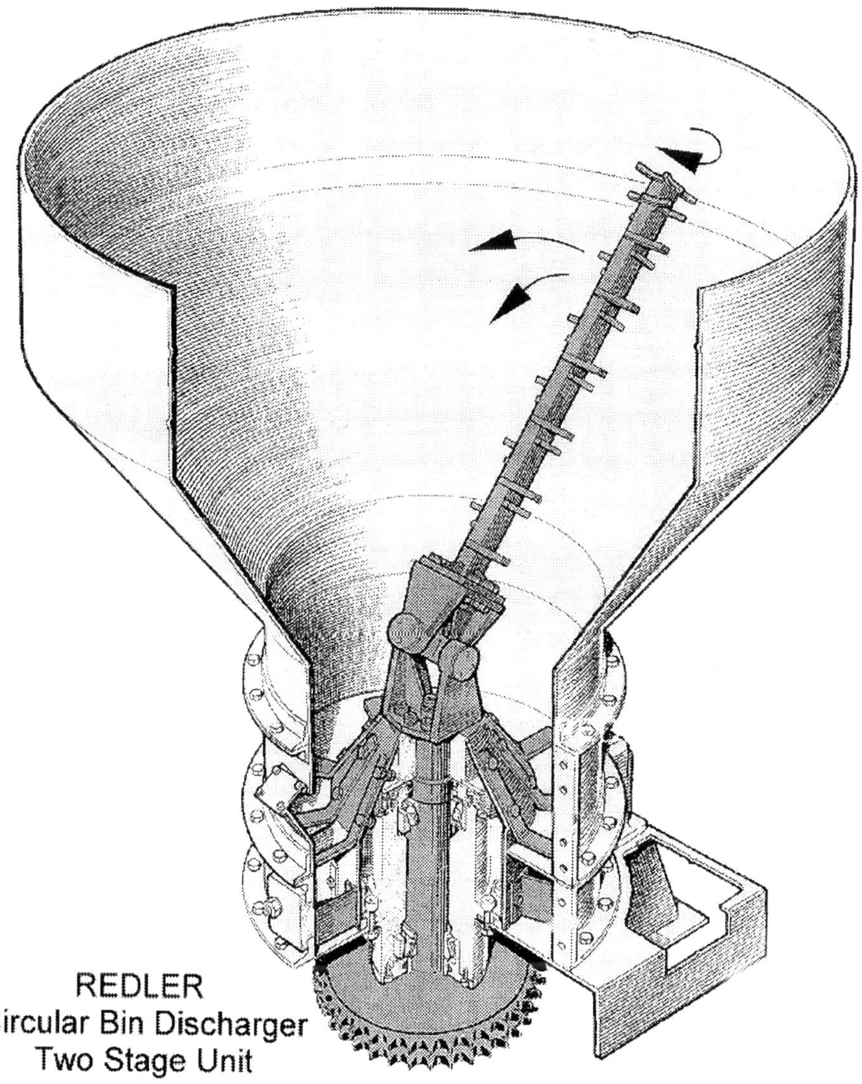

REDLER
Circular Bin Discharger
Two Stage Unit

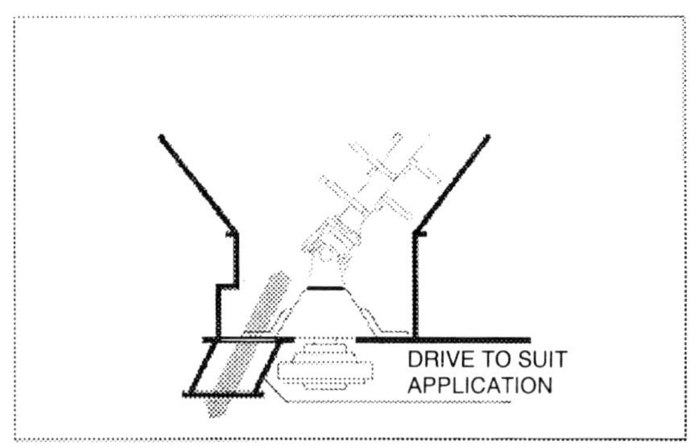

Single Stage
Circular Bin Discharger

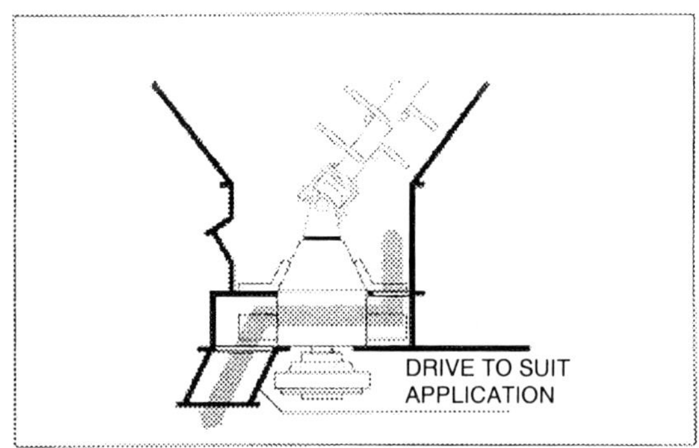

Two Stage
Circular Bin Discharger

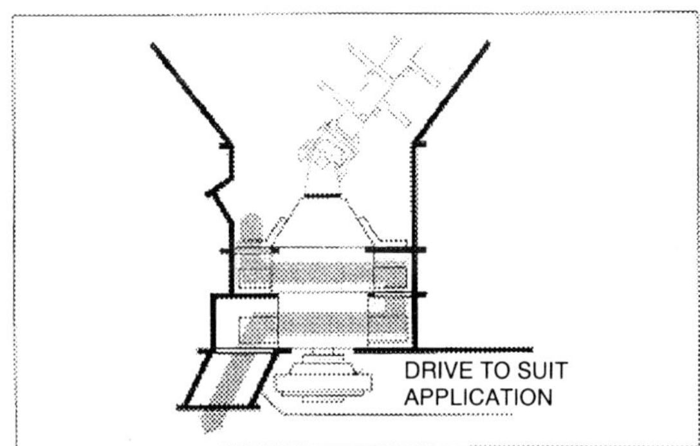

Three Stage
Circular Bin Discharger

S604/007/98

The Morillon hydrascrew planetry discharger

J E GREEN
Standprime Limited, Chelmsford, UK
L MORILLON
Morillon SA, Andreze, France

The storage of large quantities of bulk product often lead to the use of large diameter circular storage silos, a rational and enconomic solution. However whilst it is invariably an easy matter to fill such silos it is quite another to empty them especially when storing fine, cohesive, interlocking or oily products.

For such situations, and where traditional methods of steep angled hoppers, extraction by multi screw, vibratory dischargers or fluidisation prove ineffective, the Morillon HYDRASCREW planetary discharger offers a more than adequate solution.

Morillon Sa is a long established company, founded in 1865,during which time it worked continuously in the field of materials storage and handling. In the late seventies French animal feed millers were facing particular storage problems whereby there was a need for-

- Large storage capacities
- Long storage periods
- Many different materials to be stored such as oil seed meal, tapioca and various derivatives all of which were non free flowing.

Having considered the problems and pitfalls of equipment already in the market place Morillon decided, to meet these specialist requirements, to take a fresh look at the design of a reliable silo unloader with low maintenance cost and polyvalent in terms of silo size and stored material. Modern control technology along with hydraulics were used and as a result the HYDRASCREW was born.

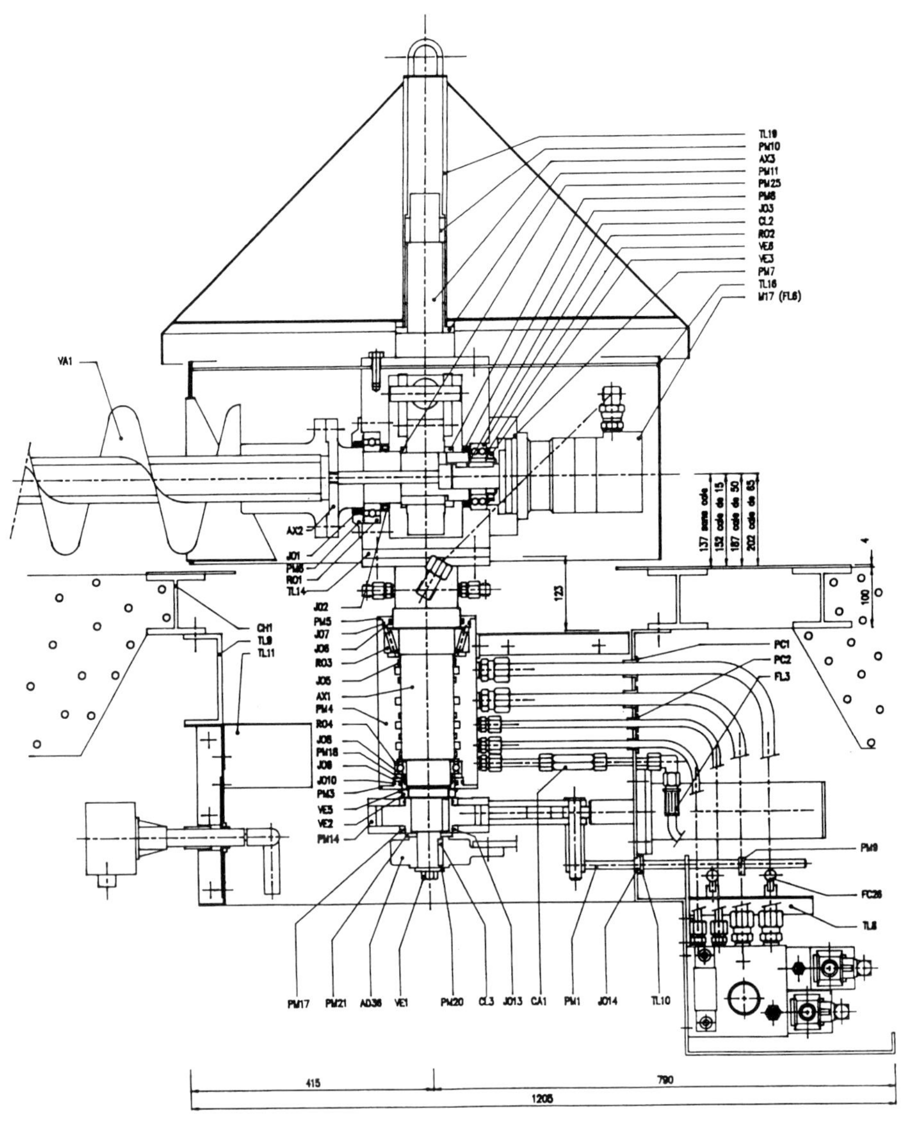

HYDRASCREW WITH PIVOT

The HYDRASCREW is designed to BOTTOM UNLOAD circular flat bottomed silos by means of an open rotating screw under the silo load. The screw revolves 360 degrees around a central support pivot (or slewing ring) and draws the product to a central outlet where the stored material is discharged to a takeaway conveyor or in the case of bulk outloading direct to bulk vehicle or IBC.

FUNCTIONING OF HYDRASCREW

HYDRASCREW is designed to safely discharge cohesive materials on a FIRST IN/FIRST OUT basis allowing the stored material to be evenly discharged down the silo avoiding BRIDGING and RAT HOLING. The cantilevered screw driven by a hydraulic motor -

- Controls the products volumetric flow.
- Creates a specific milling action to loosen compacted products.
- Ensures, by useing a screw with variable pitch and diameter, that the stored material is evenly discharged across the silo floor.
- Screw dimensions are calculated to suit the material type and required capacity.

HYDRASCREWs hydraulic system is shown on Drg R4532.

The HYDRASCREW is driven by a single hydraulic powerpack which can drive upto ten machines if only one HYDRASCREW operates at a time.

The hydraulic motor used is of Danfoss type directly coupled to the screw. As a result no intermediate gearboxes, chain or belt drive is used. The Danfoss motor is chosen because of its capability to delivery high torque at low speed. The average screw speed for HYDRASCREW is 80 rpm although this can vary depending on the material type to be discharged.

On its return to the powerpack tank the hydraulic oil used to drive the screw also drives the ratchet (or pinion) cylinder which acuates the planetary movement of the screw.

In applications where the product can prove difficult to move on commencement of discharge, or during normal operation the Morillon patented BOOSTER system is incorporated. This system detects overloading of the predetermined pressure set for the screw and causes both planetary and rotary motion of the screw to automatically stop. An auxiliary hydraulic cylinder attached to a ratchet on the screw automatically engages and causes the screw to be ratcheted around at 10 to 15 times normal torque whilst remaining in a stationary planetary position. Once the screw is freed and the pressure returns to its operational norm both planetary and rotary screw motion will automatically start or restart as the case may be. The Morillon patented BOOSTER system not only frees the screw in such conditions it also protects the screw shaft against operational abuse.

A Planetary movement controller is included in the HYDRASCREW package. This tracks against a preset time the rate of progress of the screw around the silo floor and thus ensures the silo is evenly discharged throughout its full 360 degrees. If for any reason this is not achieved an alarm is given.

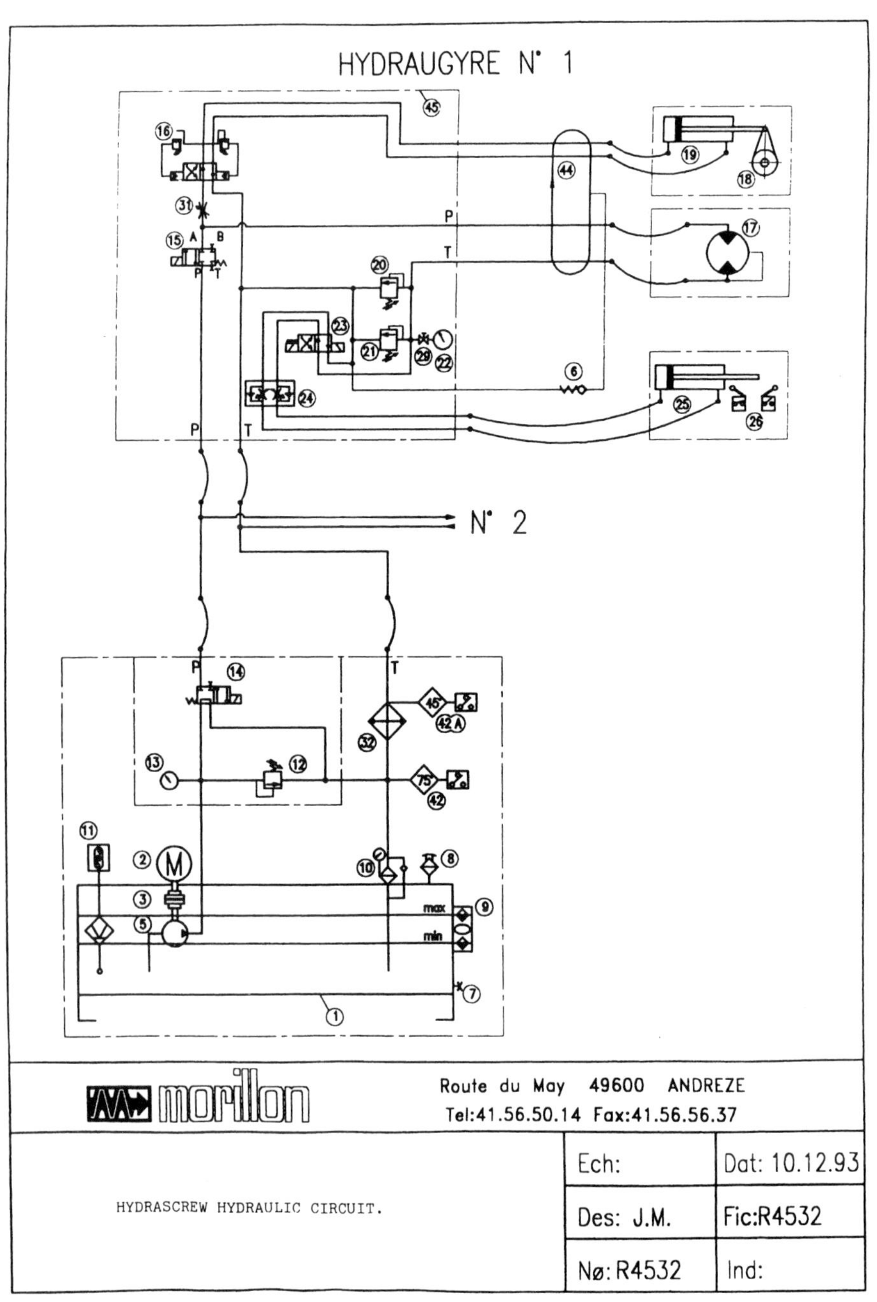
HYDRAUGYRE N° 1
N° 2
P
T
A
B
max
min
morillon
Route du May 49600 ANDREZE
Tel:41.56.50.14 Fax:41.56.56.37
Ech:
Dat: 10.12.93
HYDRASCREW HYDRAULIC CIRCUIT.
Des: J.M.
Fic:R4532
Nø: R4532
Ind:

REP	DESIGNATION	
1	Reservoir hydraulique	Oil tank
2	Moteur electrique	Electric motor
3	Accouplement	Coupling
5	Pompe	Pump
6	Clapet anti–retour	Check valve
7	Bouchon de vidange	Emptying plug
8	Bouchon de remplissage	Filling plug
9	Niveau visuel avec thermometre	Visual level with thermometer
10	Filtre retour	Return filter
11	Indicateur electrique niveau huile	Electric detector of oil level
12	Soupape de securite	Pressure relief valve
13	Manometre 0/250 bars	Manometer 0/250 bars
14	Electro–distributeur by–pass	Solenoid valve by–pass
15	E.D.ouv.ferm.circuit	S.V.for circuit opening
16	Distributeur a pilotage hydraulique	Hydraulic piloted valve
17	Moteur hydraulique	Hydraulic motor
18	Roue libre de demarrage	Booster free wheel
19	Verin de demarrage	Booster cylinder
20	Valve pression pour mouvement planetaire	Pressure valve for planetary movement
21	Soupape de securite sur mvt planetaire	Safety valve for planetary movement
22	Manometre 0/60 bars	Manometer 0/60 bars
23	Electro–distributeur pour mvt planetaire	S.V. for planetary movement
24	Etrangleur pour vitesse du mvt planetaire	Throttle valve for planetary movement speed
25	Verin pour mouvement planetaire	Cylinder for planetary advance
26	Fin de course du mvt planetaire	Limit switches
29	Isolateur de manometre	Manometer insulator
31	Etrangleur pour vitesse mvt demarrage	Throttle valve for booster speed
32	Aero–refroidisseur	Oil cooler
42	Thermostat 75°C	Thermostat 75°C
42A	Thermostat 45°C	Thermostat 45°C
44	Joint tournant	Pivot
45	Bloc fore	Hydraulic block

morillon

Route du May 49600 ANDREZE
Tel:41.56.50.14 Fax:41.56.56.37

HYDRASCREW HYDRAULIC CONTROL SYSTEM. NOMENCLATURE DU SCHEMA HYDRAULIQUE R4532	Ech:	Dat: 10.12.93
	Des: J.M.	Fic:R4532
	Nø: R4532.1	Ind:

THE TWO MAIN TYPES OF HYDRASCREW

Since its inception the HYDRASCREW has been further developed which has resulted in the application range for the machine being substantially widened. Over 1000 machines have been installed worldwide of the two available types.

a) The Pivot type. (See Drawing)

This range covers the original design of HYDRASCREW for general duty applications. The planetary drive is achieved by a hydraulic cylinder connecting to a ratchet which drives through the machines central pivot.

b) The Crown type. (See drawing No Z4945)

This range is designed for heavy duty applications and generally for larger diameter silos. The planetary drive is achieved via an external slewing ring which is engaged by a pinion wheel connected to an external hydraulic ram. This design results in a very stable design necessary for the high torque required when discharging large diameter silos. This machine has a substantially larger outlet for high volume discharge or for the discharge of light or fibrous materials. Furthermore provides improved maintenance access.

The result is the HYDRASCREW is a highly flexible machine.

The HYDRASCREW is suitable for silo diameters ranging from 2 to 20 m diameter.

Discharge rates from 5 to 400 m^3/h can be achieved.

HYDRASCREW can discharge a diverse range of materials such as pulverized coal, sludge, woodchips, oil seed meal, cement, flyash, chemicals and oxides, human and animal foodstuffs.

Silo loads as little as 500 kg of light density material and as large as 7000 tonnes of flyash have been sucessfully discharge with a single HYDRASCREW.

CONTROL AND SAFETY SYSTEM-

Powerpack

The oil level and temperature are continuously checked. If a default is detected the motor will automatically close down.

The hydraulic pressure is controlled by a pressure relief valve.

The oil flow can be controlled in order to vary the screw rpm and subsequently the machines discharge rate-

– with hand wheel attached directly to the pump.
–or with a load sensing hydraulic system piloting a proportional valve which can be controlled by a potentiometer mounted in a control panel or by the 4-20mA signal of a computor.

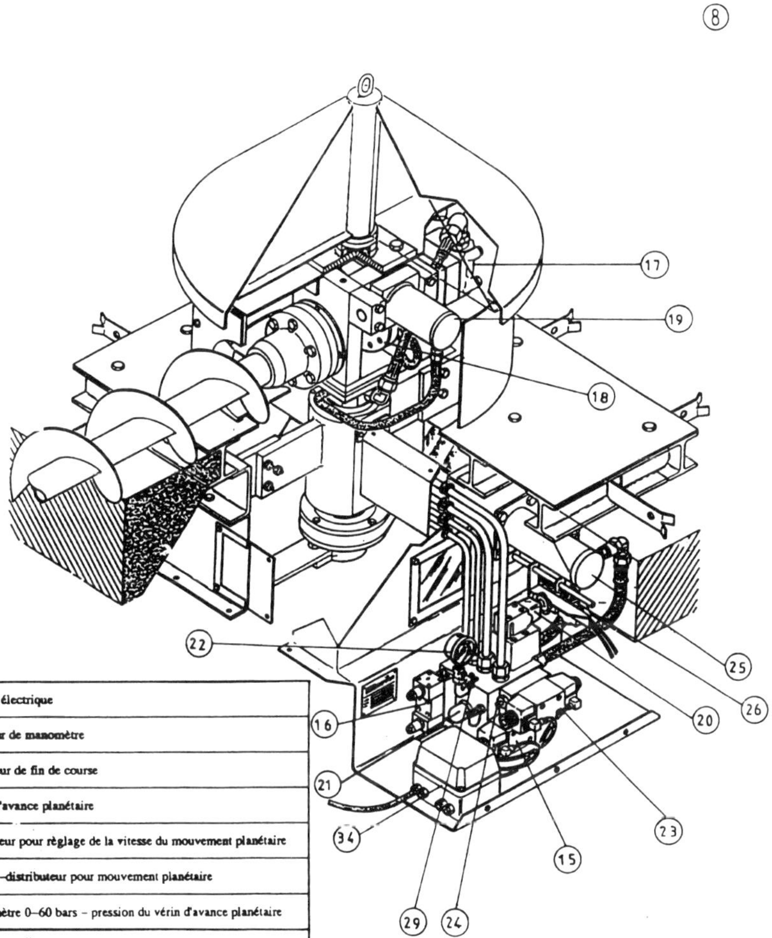

8

34	Coffret électrique
29	Isolateur de manomètre
26	Détecteur de fin de course
25	Vérin d'avance planétaire
24	Etrangleur pour règlage de la vitesse du mouvement planétaire
23	Electro–distributeur pour mouvement planétaire
22	Manomètre 0–60 bars – pression du vérin d'avance planétaire
21	Soupape de sécurité sur pression du mouvement planétaire
20	Etrangleur pour règlage de la pression du mouvement planétaire
19	Vérin de démarrage
18	Roue-libre de démarrage
17	Moteur hydraulique
16	Distributeur à pilotage hydraulique
15	Electro–distributeur ouverture de circuit

morillon — S.A. Route du May - ANDREZÉ 49600 BEAUPRÉAU

HYDRASCREW WITH PIVOT.
HM - HM1

Ech: | Date: 23-5-91 | Des: MA

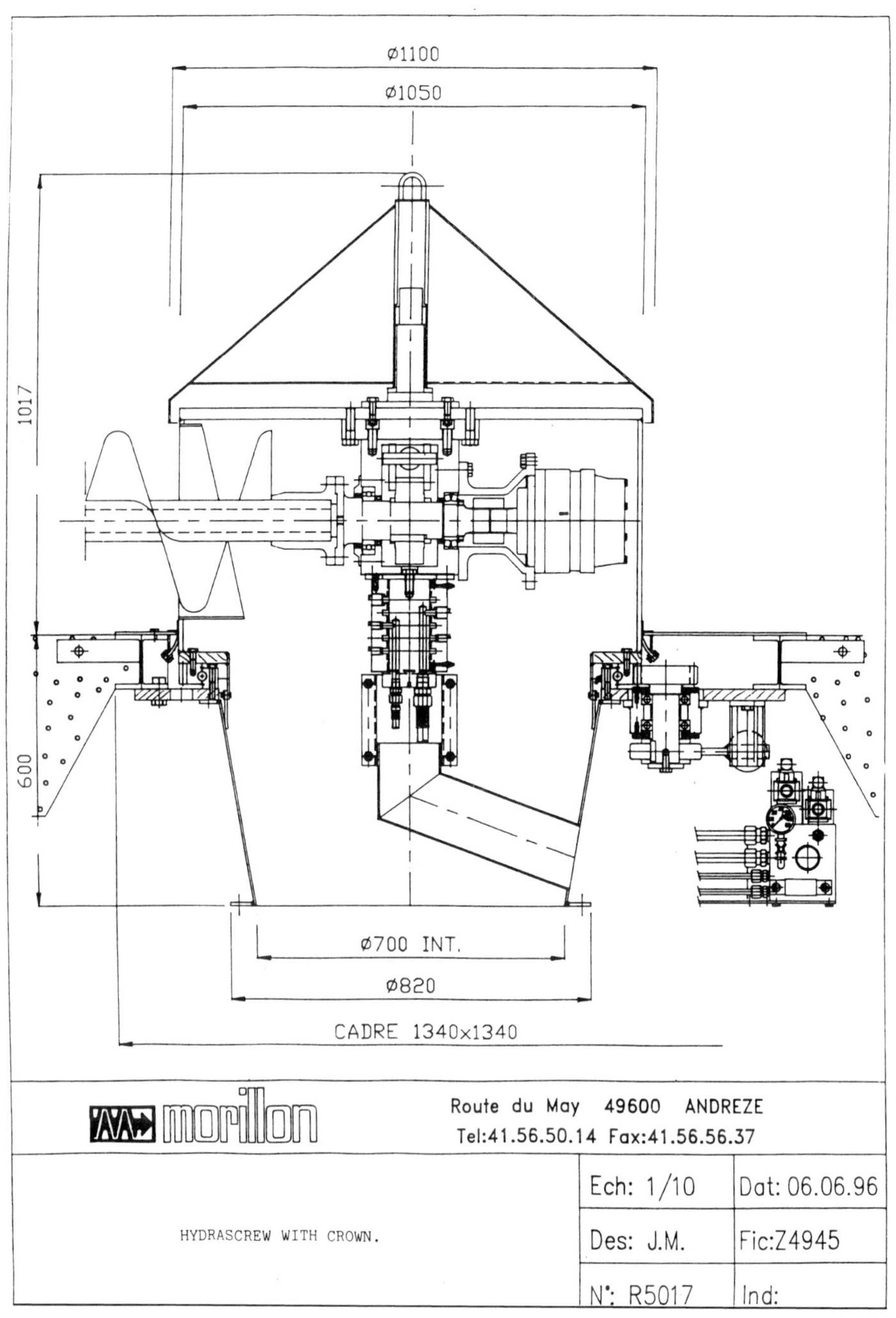
Ø1100
Ø1050
1017
600
Ø700 INT.
Ø820
CADRE 1340x1340
morillon
Route du May 49600 ANDREZE
Tel:41.56.50.14 Fax:41.56.56.37
HYDRASCREW WITH CROWN.
Ech: 1/10
Dat: 06.06.96
Des: J.M.
Fic:Z4945
N°: R5017
Ind:

Hydrascrew

The BOOSTER is controlled by a hydarulic pilot valve

The planetary movement pressure and speed are controlled by a relief valve which can be manually adjusted.

The PLANETARY MOVEMENT CONTROLLER detects any default in the planetary movement and gives a visual or audible signal to the control panel.

The total pressure of the system, indicated by a manometer on the powerpack, and the pressure required by the planetary movement, indicated by a manometer on the HYDRASCREW itself, combined with the frequency of use of the machine (cycles per day, hour or minute per cycle) can be registered and checked by MORILLON in France via phone lines useing ISDN protocol.

Models Available

The HPP -	Suitable for silos from 2.5m to 4.0m diameter. Light duty machine capable of discharge rates upto 80 cu/m hour.
The HM & HMC -	The model HM is suitable for silos upto 8.5m diameter whilst the HMC fitted with planetary drive via a slewing ring is suitable for silos upto 10 m diameter. General duty machines capable of discharge rates upto 150 cu/m hour.
The HG & HGC -	The HG and HGC machines are of similar design to the HM range but capable of discharging silos upto 13m diameter at rates upto 220 cu/m hour.
The SHG -	The SHG is capable of discharging silos upto 20 m diameter at rates upto 400 cu/m hour. The machine is supplied only with the planetary drive by slewing ring. A twin screw version with apposing screws mounted on the same central support is available. This version is for high volume discharge of light density materials.

CONVENTIONAL RECLAIMING SYSTEM

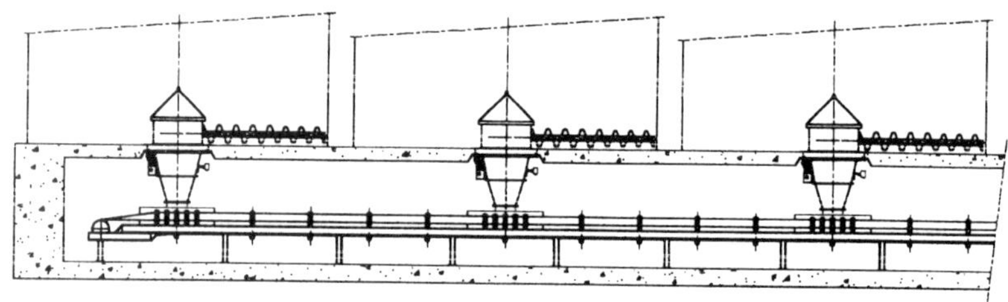

BULK VEHICLE LOADING

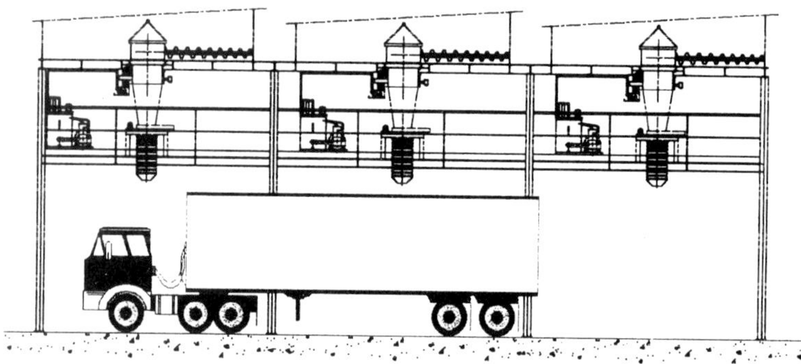

MACHINE FEEDER

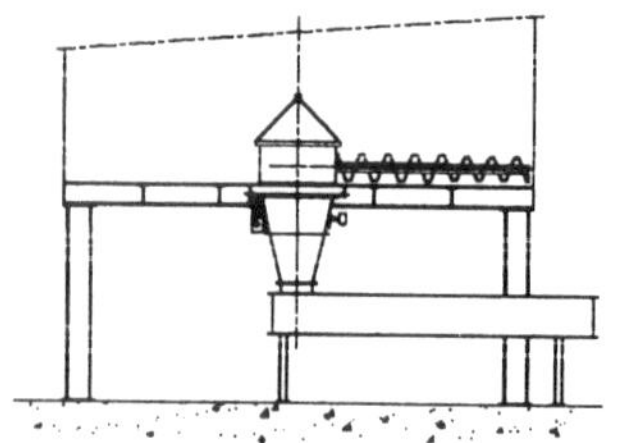

DOSING INSTALLATION

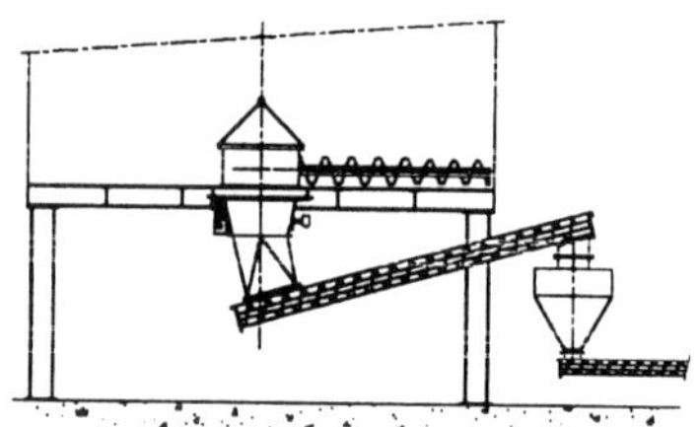

MODIFICATION OF AN EXISTING CONICAL SILO BOTTOM TO A FLAT SILO BOTTOM

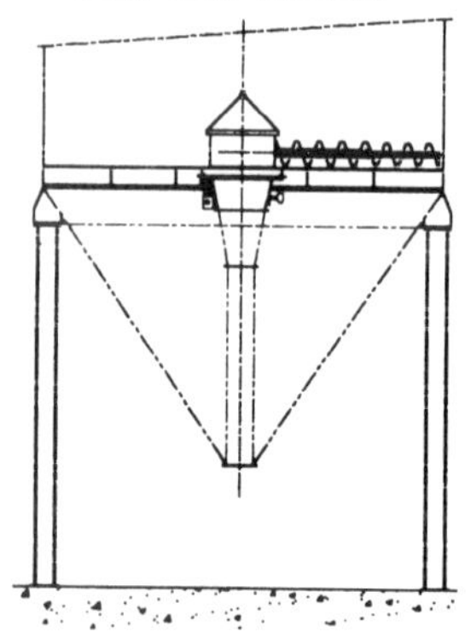

SOME OF OUR APPLICATIONS WITH HYDRAULIC SCREW UNLOADER

–HYDRASCREW–

HYDRASCREW AT RINORU OIL MILLS-JAPAN
11m diameter x 30m high silo for
SOYA BEAN MEAL.Discharge cap.200cu/m hour.

PRB FRANCE-Silo 15m diameter x 20m high

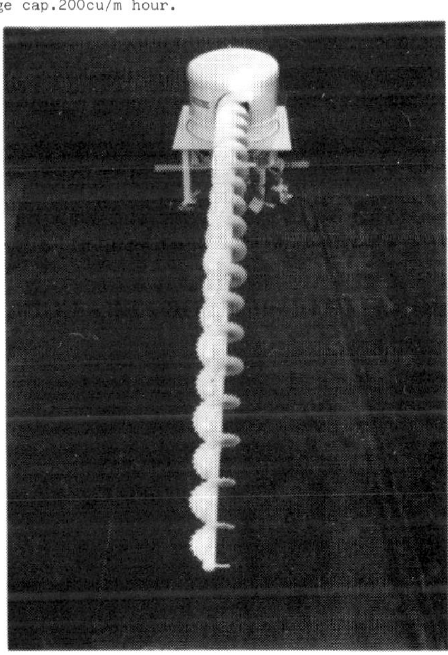

SHG HYDRASCREW AT PRB FOR CEMENT.

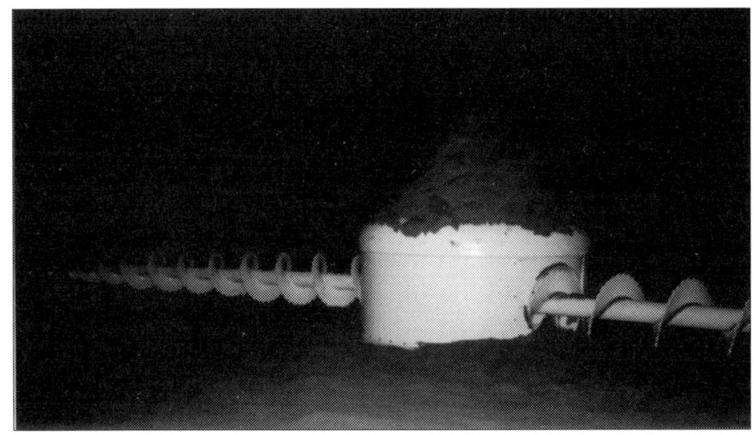

TWIN SCREW HYDRASCREW AT RINORU OIL MILLS–Japan
capacity 400 cu/m hour SOYA BEAN MEAL.

55kw powerpack for SHG HYDRASCREW.

7000 tonne FLYASH SILO AT FERRYBRIDGE
POWERSTATION with SHG HYDRASCREW.

S604/008/98

Promoting flow from silos using aeration and fluidization

J PETHULLIS
Portasilo Limited, Huntington, UK

SYNOPSIS

The use of low pressure air to partially fluidise powders in order to promote flow from silos has been established for many years and is usually the most cost effective solution to discharge problems. This paper compares the various techniques and highlights some of the pitfalls which must be avoided.

1. INTRODUCTION

The terms 'fluidisation' and 'aeration' are often interchanged and used rather loosely.

Fluidisation should, I believe, be the term applied to the introduction of air into powders, whereas the term 'aeration' should relate to the introduction of air into liquids.

Fluidisation infers that the powder will take on liquid-like properties. Whereas this can be true, if sufficient air is added to a suitable powder, it is often not necessary for successful silo discharge. This term will only be used when the bed of particles become fully supported.

More often, relatively low volumes of air are introduced into a powder to reduce its strength and improve its flow characteristics. This will be termed 'aeration' throughout this paper.

Various techniques for aeration are described in later sections. It is important first to establish if the product can be aerated or fluidised, and to determine the necessary air flow rates.

Geldart's classification of fluidisation behaviour can be used as a guide to the suitability of powders, provided the particle density and mean particle size is known. This classification is reproduced in Figure 1. Powders in categories A & B are usually very suitable for aeration assisted discharge from silos using techniques described in Section 2 or 3. Powders in Category C tend to be very cohesive and difficult to fluidise, but can be discharged quite successfully with a modified design described in Section 4. Powders and granules in Category D tend to need very high air flows to fluidise and an alternative discharge mechanism is often preferable.

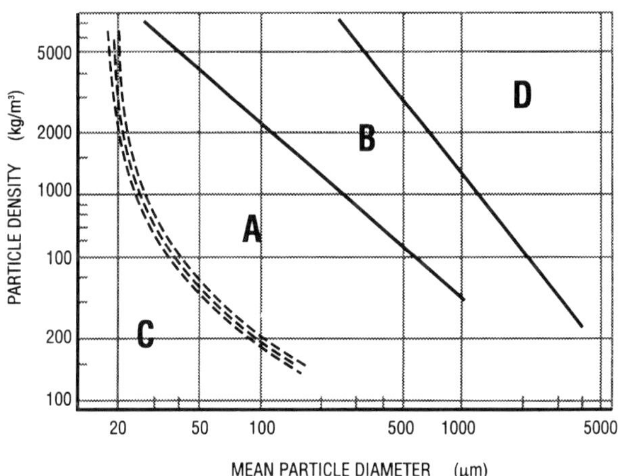

Fig 1 Geldart's classification of fluidisation behaviour according to size and density of the particulate material.

In practice, it is often easier to measure directly, the air flow rate and powder permeability in a small scale laboratory test rig, than it is to measure the powder particle size and density.

Jenike (Ref 1) describes an ideal air-hopper, which incorporates air panels over the entire hopper surface. Air addition is used to reduce the angle of friction at the walls close to zero and fluidise the powder only in the region of the hopper outlet.

The optimum slope angle for a conical device is around 45° to the horizontal. By 'lubricating' the hopper walls in this way and fluidising only the hopper's outlet section to prevent bridging, air consumption is minimised. The hopper does not rely on an expanded bed, whereby all the air flows upwards through the silo. Air flows out of the hopper through the outlet. This configuration is therefore, best suited to discharge into airslide type conveyors and described fully in Section 2.

In can be shown that hopper angles in excess of the ideal air hopper lead to reduced flowability, and that it is therefore, undesirable to place air panels on steep hoppers.

In practice, partial aeration using a limited number of air pads is often combined quite successfully, with a relatively steep (typically 60°) hopper. The purpose of the pads is to weaken the inter-particle bonds and prevent bridging. This solution is described in Section 3.

Equally, in the case of highly cohesive and difficult to fluidise powders, steep hoppers fitted with a porous membrane over the entire surface area can be successful when a hopper of 'ideal' proportions will not discharge at all. This is described more fully in Section 4.

2. 'IDEAL' CONFIGURATION FOR SILO AERATION

Many products which aerate readily are handled in silos with a hopper bottom geometry approaching the ideal as described in the preceding section. Typical examples of these products would be Cement, Pulverized Fuel Ash and Alumina.

Because the silos are often combined with airslide conveyors, and because airslide conveyors require significant headroom due to their ability to convey uphill, the hopper angle is often reduced. This may mean that additional air is required, but this is not significant when considering the overall capital costs.

In addition, due to the ease with which these materials can be aerated, or even fluidised, it is common for air pads to be fitted to only part of the hopper surface.

Such an arrangement is shown in Figure 2, which shows the hopper section of a large PFA bunker. The hopper angle in this case has been reduced to 15° to the horizontal.

Fig 2

When making such a compromise on hopper angle, consideration should be given to the likely side effects. Core flow can certainly be expected, which in itself may not be a problem. However, there can be a significant loss in live storage capacity and stagnant non-flowing zones can eventually consolidate and become a permanent feature of the bin. This is not acceptable if the product has a limited shelf life.

Core flow also promotes segregation, and severe problems have been reported in large Alumina bins due to this phenomena. In this particular case, variation in particle size distribution did not effect the process, but badly affected the performance of downstream airslide conveyors.

3. AIR PADS TO ASSIST GRAVITY FLOW

Despite the fact that the optimum air-hopper would have a 45° angle to the horizontal and have air panels over the complete surface area, perhaps the commonest application of aeration seen in the industry is a relatively steep hopper, typically 60° fitted with a number of local aeration panels. Such an example is shown in Figure 3.

By increasing the hopper angle and utilising gravity flow, the coverage of the air panels can be reduced substantially.

Although not ideal, the compromise is workable and this design will typically be the lowest cost solution to a silo discharge problem.

Fig 3

Bridging is usually perceived as the flow problem, and is quite easily dealt with by fluidising the powder below the critical bridging diameter. The bridging diameter can be calculated from a knowledge of the product's shear properties, but this will tend to be conservative, since the properties are measured in a de-aerated sample. Judgement can be exercised here since the cost of an extra row of pads is not high and it is relatively easy to extend the fluidisation zone.

Typically, the air pads are capable of passing between 2-10m^3/hour of air with an air supply pressure of 0.7 bar gauge. The number of pads required is varied depending on the product to be handled, the silo size, the outlet size and, often overlooked, the discharge rate. Guidance will usually be given by the supplier of the air pad as to the number of pads required, but this will usually only relate to prevention of bridging.

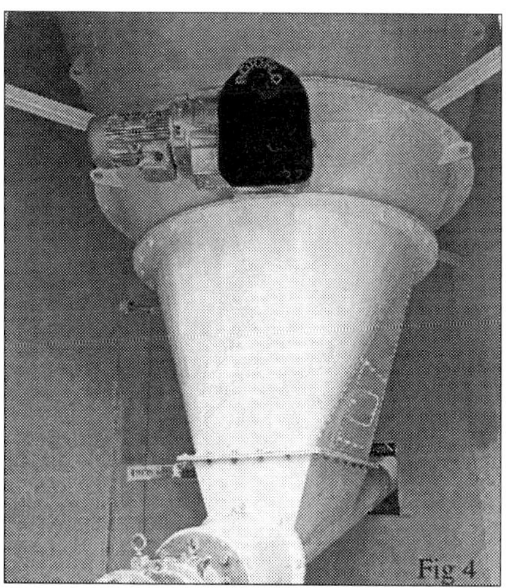

Fig 4

Concentrating the air pads around the silo outlet will encourage core flow at best, and piping at worst. The central column of powder within the silo, having a diameter approximately equal to the bridging diameter, can become fluidised. In extreme cases, this will discharge completely with very little other powder being discharged. Air pads installed further up the hopper allow latitude in estimation of the critical bridging diameter, but do little to alleviate the problem of stagnant zones. It is interesting to note that a further increase in hopper angle will also do little to move this powder, even though the hopper angle attained would be sufficient to generate mass flow were it combined with the correct sized outlet for unassisted gravity flow.

Where core flow or segregation are undesirable, more extensive aeration systems as described in Sections 2 or 4 should be considered, or an alternative discharge device should be used. Figure 4 shows a Portasilo ROTOFLO discharge valve combined with a small amount of aeration and a 70° hopper angle. These are used together to generate mass flow in a 100 tonne Titanium Dioxide storage silo. The small quantity of air allows for a significant reduction in machine size and power.

4. MODIFIED DESIGN FOR VERY COHESIVE POWDERS

The ideal air-hopper geometry cannot be utilised with highly cohesive powders such as Titanium Dioxide, Zinc Oxide and very fine grades of Calcium Carbonate for example. These products are very difficult to fluidise, the air forms into channels at high air flow rates and does not penetrate the powder bulk.

By increasing the hopper angle to 60° or more and aerating the entire hopper surface area, it is possible to fluidise a thin layer at the hopper boundary. Air flow rates are very dependant on the product characteristics and pilot scale tests are essential.

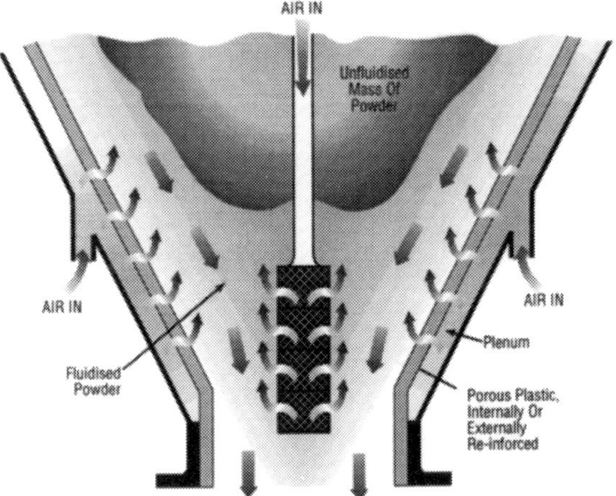

Fig 5 Air Lance For Difficult To Fluidise Products

On hoppers with a large surface area, it may be necessary to partition the plenum and regulate the flow of air to each section to avoid preferential flow in the upper regions.

Having introduced air to lubricate the hopper walls, it is essential that the powder in the outlet region is fluidised or no air will penetrate the bulk. Air is best introduced in this region by a lance as shown in Figure 5. Air pipework can enter from above in small silos, or below or to one side in the case of larger installations.

5. AIR PAD DESIGN

A number of small air pads, as described in Section 3, are available from different manufacturers. These can be divided into three groups as follows:-

a) Internally Mounted Diffusers.

Typically manufactured from sintered plastic or metals, these are designed to be as low profile as possible, but can tend to obstruct product flow. Ease of fitment is the key benefit, requiring only a hole drilled through the hopper wall. Access into the silo is required to change pads, which can be a major drawback.

b) Externally Mounted Diffusers as Shown in Figure 2.

Again, available in sintered plastic or sintered metal for high temperature applications. These fit flush on the hopper wall and can be changed from outside the vessel. They are not so easy to retrofit as the steel boss requires welding to the hopper wall.

c) Internally Mounted Devices with Rubber Boots.

In this case, the diffuser is replaced by a rubber boot, which prevents back-flow of fluidised powder when the air is switched off. Additionally, the rubber boot is said to vibrate when air is applied for additional agitation. This can tend to obstruct the product flow and needs access into the silo to fit and change.

Larger air panels are usually fabricated with plastic or fabric diffusers and can be fitted flush with external plenums, or internal to the hopper.

As with the smaller pads, the external design is preferable for maintenance and to minimise obstructions to flow, but structural reinforcement of the hopper cut out is required. This can be significant when large pads are fitted to large vessels.

Lining of the full hopper surface area is usually carried out using porous plastic or fabric membranes. The mechanical strength of both of these is limited and some reinforcement is required. This can take the form of a perforated mesh sandwich construction as shown in Figure 6. Here flat panels are being laid on a grid of supporting steelwork, which forms the air plenum. The porous plastic sheet is also available with internal reinforcement, which simplifies construction and has a smooth outer surface, which is very beneficial to flow.

Fig 6

On smaller vessels, it is possible to fabricate complete conical hopper from the porous plastic, with formed flanges for attachment to the vessel.

Fabric membranes tend to be lower cost, but care must be taken in selecting the right fabric. Too coarse a cloth can lead to powder penetrating the weave and rapid blinding of the pores. Nomex and Kevlar fabrics are available for high temperature applications.

6. AIR DISTRIBUTION

Correct distribution is essential to prevent the air short-circuiting and taking the easier route to atmosphere.

Small systems, as described in Section 2 tend to be supplied with air via small bore nylon tubing. Here, the pressure drop through the tubing tends to exceed the pressure drop through the pad and the powder. Distribution manifolds are therefore, required to prevent the majority of the air passing through the first pads in the system.

On larger systems, the air is typically supplied through a ring main with a relatively low pressure drop. This leads to even distribution between different pads on a row, although there will still be a tendency for air to flow through the upper rows of multi-row systems as the silo empties. This effect is best countered by selecting a diffuser with a low permeability or using orifice plates or valves to restrict the flow to the upper rows of pads. A higher pressure will be required to the upper rows of pads. A higher pressure will be required at the blower to overcome this increased resistance.

7. AIR SUPPLY SYSTEM

The volume of air will be calculated from the type and number of pads or panels fitted. The pressure drop is estimated from a summation of the losses in the distribution pipework, across the membrane and through the powder column in the silo.

Whereas the first two elements are readily calculated, the latter is usually best measured directly using test equipment. Care should be taken when scaling from small scale tests since permeability is not linear in a powder column. Figure 6 shows a large scale test column used by Portasilo to measure permeability of aerated and fluidised columns. Lack of linearity is due to channelling of the air and preferential flow at the silo walls.

Provided distribution losses are minimised by correct sizing of the pipework, the total back-pressure for most applications can be kept below 0.3 bar gauge, allowing the use of side channel blowers. These are not positive displacement machines and are therefore, more forgiving of slight mis-calculation in pressure drop. On small systems where low cost small bore tubing is used for distribution of the air, high pressures are typically encountered and sliding vane or diaphragm blowers operating at 0.7-1 bar gauge are usually preferred.

Whichever blower is selected, a non-return valve should always be fitted in the discharge side. When the blower is stopped, the silo will de-pressurise back through the air pads if such a device is not in place. Minor leakage of a failed diffuser or diffuser seal will destroy the blower.

For the majority of powders, drying of the air is unnecessary. A typical design figure for United Kingdom ambient conditions is an absolute humidity of 40 gms per kg of air. Provided the blower outlet temperature exceeds 25°C, then the relatively humidity of the air entering the silo will be below 60% RH.

If the product is hygroscopic and particularly sensitive to moisture, the equilibrium humidity of the powder should be measured. If the blower discharge conditions can exceed this figure, dehumidification of the air is required or an alternative discharger should be considered.

Compressed air from a works air main is often used for smaller systems, the pressure being controlled with a locally mounted regulator. Here, the air will not be warm and some form of drying is recommended if not already incorporated after the compressor. The dryer should be capable of achieving a pressure dew point below the anticipated minimum ambient temperature, which usually requires desiccant type dryers.

Porous plastic diffusers have very small pores, which give good resistance to blinding from the powder side. They are however, very sensitive to blinding from the air side, if the air is not adequately filtered at the pressure regulator or the blower intake.

8. SUMMARY

Various designs of aeration system have been discussed, which can give trouble-free discharge at relatively low cost if correctly selected. Premature failure of the air pads or panels can usually be attributed to poor air quality, so this detail should not be overlooked.

Major problems likely to be encountered with aeration systems are flooding and stagnant zones within the silo.

The potential of flooding must never be under-estimated, because the result can be catastrophic.

Some form of feeder is usually incorporated below the silo and this must be selected with containment of a fluid in mind. Screw feeders can be over-run, especially if short or horizontal and provision of a separate on-off valve at the feeder outlet must be considered. This same comment applies to drag conveyors. Belt conveyors should not be considered without a rotary valve on the silo outlet. It should be noted that if the product is fluidised, hydrostatic pressures at the silo outlet can cause damage to flat conveyor covers, transition chutes or the like if not adequately designed.

A great deal has already been said about the likelihood of core flow and the problem of non-flowing stagnant zones. The live capacity of small silos can be reduced significantly, to the point where a full tanker load cannot be accommodated when refilling. In this case, or when the product has a limited shelf life, an alternative discharger should be considered.

In conclusion, I cannot over-emphasise the importance of paying careful consideration to the characteristics of the product for each specific application. A lack of attention at the design phase can lead to an on-going problem for plant management and a waste of time, product and therefore, money.

REFERENCES

1. A J Jenike
'Storage and Flow of Solids, Bulletin No 123', Utah Engineering Experiment Station

S604/010/98

Low-friction and low-adhesion linings

J HILLS
3 J Linings Systems Limited, Doncaster, UK

1 INTRODUCTION

The use of ultra high molecular weight polyethylene (UHMW) plastic liners to overcome flow problems encountered in bunkers, silos and chutes has been successfully used for many years.

This has been most significant in handling problems encountered on existing plant, particularly where there have been changes in the type and nature of the product being handled. On new plant it is obviously possible to design for mass flow and optimum performance. In that case UHMW linings can offer significant advantages in flow performance and also in cost effective design of bunkers and silos. By using plastic linings, overall heights, angles of cones and outlet size can be reduced.

However in the majority of cases, problems are encountered on existing plant where the original product has been changed. These problems can be solved by the use of lining materials to improve the flow at the bunker wall.

These include:-

- Ultra High Molecular Weight Polyethylene
- Glass
- Stainless Steel

The selected lining system will depend upon a wide range of factors:-

- Type of product handled
- Product size
- Product density
- Abrasiveness of product
- Moisture content of product
- Chemical properties - including corrosive nature
- Expected Life
- Cost

The most significant use of UHMW has been in areas where the properties of the material can be used to best advantage.

1.1 Properties of UHMW

- High abrasion resistance
- Very low coefficient of friction
- Nil absorption rate

These properties make it particularly suitable for overcoming sticking and flow problems encountered in bunkers, silos and chutes. In addition it offers other advantages in the terms of operation.

These are:-

- It is simple to install
- Relatively low cost
- Needs no power and no moving parts
- Can convert core flow bunkers to mass flow

It is therefore very cost effective.

The areas where the material has been most successful are in the handling of high moisture content material which has a large proportion of "fines" content.

Crushed Coal -
Power Stations
Industrial Boiler Systems
Coal Production
Coal Transportation

Sand -
Foundry and Construction

1.2 Changes In Product

Within the minerals handling industry the increased use of water to reduce dust and mechanical methods of production have changed the product being handled. There has also been an increase in the use of imported opencast coal which has a high moisture content and a large percentage of "fines".

This change in the nature of the product results in severe handling problems. Generally it has been found that the product change results in :-

a Failure of material to flow.

b Existing handling equipment fails to perform.

When these problems are encountered plant managers and operatives normally resort to the traditional methods of improving mass flow. These are the use of hammers, rodding poles and shear brute force resulting in increased labour, and the obvious increases in running costs. (See Figure 1.)

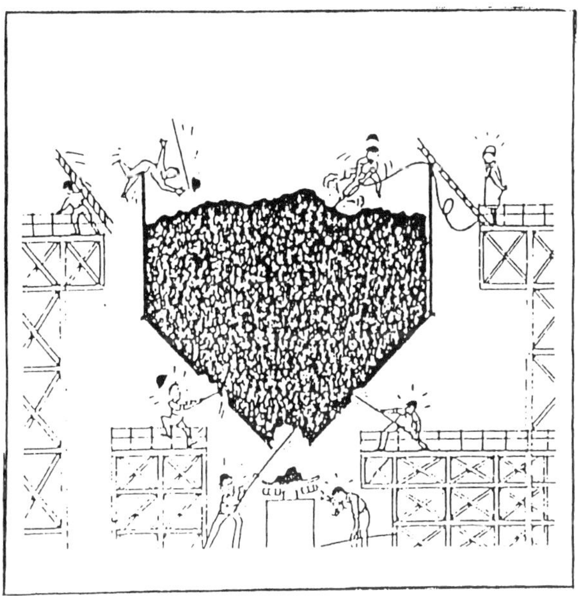

Figure 1

2 DESIGN AND SPECIFICATION OF UHMW POLYETHYLENE

2.1 What is UHMW Polyethylene?

UHMW is one of the worlds toughest known thermoplastics. It has remarkable properties including very low coefficient of friction, high abrasion resistance and nil absorption rate. It has sliding properties far in excess of polished steel and has exceptional wear properties. It has been shown that it will outlast steel many times in lining applications. However, care must be taken to avoid impact.

Compared with other plastics, UHMW polyethylene offers significant advantages and must not be confused with lower grades of materials which include high density polyethylene, low density polyethylene, PVC's, polypropylene, polyurethanes etc.

2.2 How is the Material Manufactured?

The main method of manufacturing this material is:-

Sinter Press Sheet

This involves the sintering of the basic polymer in an enclosed platen press at high temperatures and pressures. Sheets are produced in a range of thicknesses up to 200mm.

2.3 What are its Main Properties?

- High Abrasion Resistance, providing excellent sliding surface.
- Low coefficient of friction giving non-stick properties.
- Nil absorption rate, repels water.
- Lightweight, chemical resistance, unaffected by inorganic chemicals up to 95 C.
- Reduces noise.
- Improves flow rates.
- Reduces maintenance and labour costs.

2.4 What are its Limitations?

UHMW has good resistance to sliding abrasion. However, impact of material being handled should be avoided i.e. where a relatively thin lining material is fixed to a metal or concrete substrate, materials impacting will abrade the surface.

This can be overcome by correct design including the use of thicker materials, deflector plates, correct use of level indicators or provision of alternative materials for impact resistance such as polyurethane, rubber or ceramic tiles.

Due to its non-stick properties the material has to be mechanically fixed to other components. It can not be successfully bonded or glued.

2.5 Why Does It Work In Bunker And Chute Lining Applications?

As stated this material has been very successful in overcoming flow problems, particularly with high moisture content materials. Its properties of high abrasion resistance, low coefficient of friction and nil absorption rate emphasises the flow characteristics of high moisture content materials.

It has long been considered that improvement in flow is a function of the coefficient of friction of the material and is related to the angle of friction. (Table 1 shows the comparisons with other materials)

It can be seen that UHMW has a slightly higher coefficient of friction than glass, but a far better coefficient of friction than mild steel or concrete.

The resistance to moisture of the material has the effect of reducing the adhesive force and contact on the bunker wall. (See Figure 2)

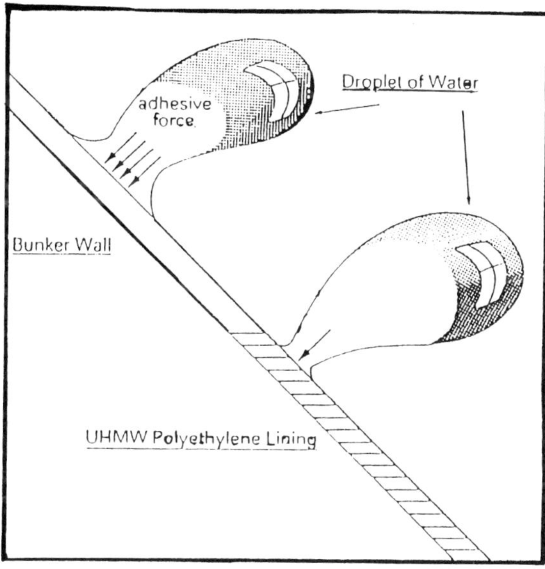

Figure 2

In major bunker applications it has been found that UHMW linings have proved successful where other systems have failed. In particular glass tile linings, whilst technically having a lower coefficient of friction, has been found to be ineffective when handling high moisture material.

Table 1
Friction Comparisons

MATERIAL	ANGLE OF FRICTION (In Degrees)	COEFFICIENT OF FRICTION
Glass	14	0.258
UHMW Polyethylene	15	0.275
Polypropylene	17	0.305
Nylon	19	0.344
Stainless Steel	21	0.373
Mild Steel	25	0.455
Polyurethane	25	0.462
Concrete	28	0.532

2.6 Sliding Properties

Table 2 gives comparisons of the sliding properties of UHMW polyethylene when compared with other materials. These figures produced by the raw material manufacturer, Hoechst, clearly illustrate the advantage of UHMW, particularly when wet.

It can be seen that UHMW polyethylene is 100% better than polished mild steel. The only material that has a technically higher sliding property is PTFE. However, this material has limitations on mechanical strength and cannot be recommended for normal minerals handling applications particularly due to its very high cost.

Work has been carried out on examining the effects of surface finish on flow properties, particularly this has been referred to in the use of polished stainless steel. However, in practise it has been found that surface finish will not have an effect on the flow characteristics encountered in bunkers and silos. This is due to the fact that the surface finish is rapidly altered in operation due to the handling of abrasive products.

Table 2
Comparison of Dynamic Coefficientof Friction on Polished Steel

LUBRICANT	Dry	Water	Oil
Mild Steel	.25 - .35	N/A	N/A
UHMW	.10 - .22	.05 - .10	.05 - .08
Nylon	.15 - .10	.14 - .19	.02 - .11
Acetals	.15 - .35	.10 - .20	.05 - .10
Nylon MOS	.12 - .20	.10 - .12	.08 - .10
PTFE	.40 - .25	.04 - .08	.04 - .05

2.7 Resistance To Abrasion

UHMW polyethylene has good sliding abrasion resistance. However, as stated it is not suitable for high abrasive materials and areas where impact occurs. For these areas alternative materials such as polyurethane, rubber, steel, basalt or ceramics should be used.

Its sliding abrasion is shown in Table 3. These figures are taken from tests carried out in laboratory conditions by Hoechst and show the relative volumetric loss measure on a sand slurry test and compare UHMW polyethylene with alternative materials.

Table 3
Sliding Abrasion

MATERIAL	SG	VOLUME LOSS
UHMW -virgin grade 1000	0.94	100
Carbon Steel	7.45	160
Nylon	1.15	210
PTFE	2.26	530
Stainless Steel	7.85	550
HDPE	0.92	600
Polypropylene	0.90	660
PVC	1.33	920
Epoxy Resin	1.53	3400

2.8 Material Specification

On bunker lining applications it is essential that the correct grade and thickness of lining material is used relative to the type of product being handled.

3 EFFECTS OF LININGS ON MASS FLOW DESIGN

Considerable work has been carried out in the design of bunkers and silos for mass flow. Papers produced by Johanson, Walker and Wright go to considerable length in setting out design criteria to ensue mass flow.

Mass Flow can be defined as a situation where all material in the bunker or silo flows together thus ensuring complete emptying of the bunker with the elimination of ratholing and build up. Mass flow is also important where segregation is undesirable and the need for complete emptying to ensure non-contamination with other products. Mass flow also ensures complete use of the storage capacity.

The alternative to mass flow is **core flow**. This is where the material flows down a central core leaving some material on the walls of the bunker. This may be desirable where stresses or abrasion on the bunker wall need to be avoided.

When designing new equipment or putting forward proposals for existing equipment, it is desirable to carry out a range of tests to ensure that the correct criteria are established. From these tests it is possible to predict specific conditions:-

- Critical arching and ratholing dimensions.
- Limiting hopper angles.

- The effects of moisture content, temperature, humidity, binders and time in storage.
- Flow aids and over pressures.
- The influence of variable particle size.

To establish these parameters it is necessary to test the material being handled to determine the strength and flow characteristics in the likely worst conditions to occur in practise. It is then possible to calculate wall pressures and final structural design surveying practical effects of UHMW lining. Where UHMW plastic linings are installed, dramatic effects on the ability of the bunker or silo to mass flow can take place.

The provision of a low friction lining to the hopper of the bunker or silo greatly increases the pressures on the structure, particularly at the transition point between upper cone and vertical walls. Figure 3 indicates the change in pressure measured on a bunker structure.

It is therefore extremely important to take into consideration the effects of these changes when lining with UHMW polyethylene. This is particularly true when lining old existing equipment. If the original equipment has been designed for core flow then stresses at the hip joint could cause failure.

It is possible, however to accommodate this problem when installing the lining material. Normally to achieve mass flow it is necessary to line the whole cone section of the bunker or silo taking the lining to the hip joint. However, where problems of possible stress may be encountered mass flow should be avoided. To improve the flow it has been found that by lining only part of the cone section of the equipment then a 'halfway house' between core flow and mass flow can be achieved. The lining of part of the cone will result in the material being held at the hip joint (Figure 3), which has the effect of spreading the pressure on the bunker wall over a greater area.

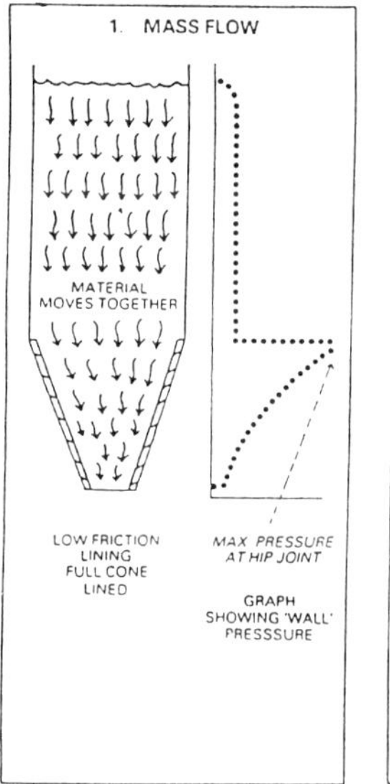

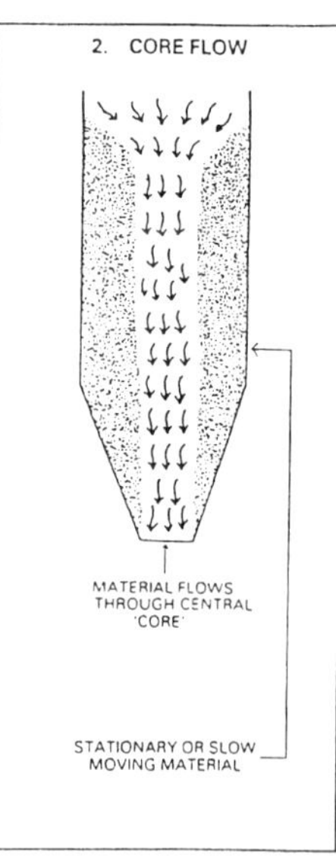

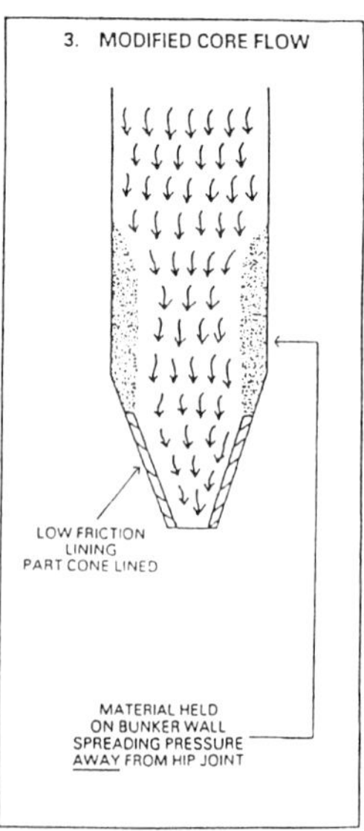

Figure 3

4 INSTALLATION METHODS

Failure often occurs due to poor installation. This results in the plastic lining being condemned by the user. However, failure can be attributed to several factors :-

- Incorrect type of fixing.
- Corrosion of fixings.
- Insufficient fixing.
- Centres to large between fixings.
- Failure to fit capping strip and valley angles.
- Incorrect jointing allowing material to ingress behind sheet.

To overcome this an installation specification has been established and this sets out the procedures to be carried out when lining.

These should include the following :-

a Correct choice of material which should be tested accordingly.

b Correct thickness of material for the product to be handled.

c Sheets laid with the longest length in the direction of the flow.

d All horizontal joints to be overlapped to a minimum of 75mm in the direction of flow.

e All vertical joints to be butt jointed.

f Sheets to be laid in brick fashion to eliminate vertical joints in the same plane.

g Top edge of lining to be capped using a stainless steel pre-formed strip to eliminate ingress of material.

h Valley angles to be fitted with radiused pre-formed sections from the same material thickness as the bunker lining.

i **Fixings**:-
Steel Bunkers. For steel bunkers and silos a range of fixings are available. The 3J Thread Forming Fixing has been specially designed to provide optimum performance in the widest range of conditions. This fixing is normally an M8 bolt with an M10 head, the bolt being thread forming, cadmium plated and PTFE coated. This fixing is also available in stainless steel. This fixing can be installed from the inside of the bunker without need for exterior access.

Concrete Bunkers. For concrete bunkers the fixing normally specified are thread forming fixings or stainless steel countersunk screws fitted with alloy anchors or plastic plugs to the concrete wall.

5 DESIGN OF BUNKERS AND CHUTES

Failure of a lining can often be attributed to poor design of the original bunkers. Poor design includes:-

- Too shallow an angle of the cone.
- Internal obstructions such as reinforcing beams.
- Outlet too small.
- Poor control of contents level resulting in bunker emptying and material impacting on the wall.
- Poor design of filling mechanism - again resulting in impact.

Where structural alterations cannot be made then the aforementioned problems can be overcome when specifying the lining :-

- Elimination of impact areas ensuring chutes and conveyor do not discharge onto bunker wall.
- Ensuring bunker does not empty completely..
- Using reliable level gauges to ensure that material impacts on itself.
- Use of deflection plates - sacrificial pieces or alternative materials more resistant to impact e.g. polyurethane, rubber or ceramic.

6 CONCLUSION

Correct use of UHMW polyethylene in the materials handling industry can go a long way to ensuring free flowing materials at an economic cost. This can be achieved by the use of the correct specification of materials and installation.

S604/011/98

The use of inserts in hoppers

L BATES
Ajax Equipment Limited, Bolton, UK

WHY FIT INSERTS ?

Hopper inserts are fitted for many different reasons, as Table 1. They form part of the tools available to the designer to optimise the performance of bulk storage containers. The first stage of the design process is to determine a form of flow regime appropriate to the physical nature of the bulk material and the circumstances of the application. A hopper of simple geometric construction, such as a cone or wedge shape, often meets the requirements of the flow channel, but other shapes, with or without inserts, may present a more economic alternative, or offer benefits that are not given by more conventional hopper forms. In some cases these inserts perform dual or multiple functions

Table 1. Reasons for Fitting Hopper Inserts

In the hopper outlet region

- To aid the commencement of flow
- To secure reliable flow through smaller outlets
- To increase flow rates
- To improve the consistency of density of the discharged material
- To secure mass flow at reduced wall inclinations
- To expand the flow channel
- To improve the extraction pattern
- To reduce overpressures on feeders
- To save headroom/ secure more storage capacity
- To counter segregation
- To blend the contents on discharge
- To improve counter current gas flow distribution
- To prevent blockages by lumps or agglomerates

In the body of the hopper

- To accelerate the de-aeration of dilated bulk material
- To reduce compaction pressures
- To alter the flow pattern

At the hopper inlet

- To reduce segregation
- To reduce particle attrition
- To secure a higher fill level
- To divert oversized product
- To counter excessive wear
- For personnel safety

FLOW REGIMES

A flow channel for a bulk material is characterized by its boundary conditions and the nature of its change in cross section. Within a bulk storage container the boundary is formed by a bed of static product or by a wall surface of the container. The resistance to sliding is invariably less on a wall boundary surface than the internal friction of a static bed interface, otherwise the material would shear within the bed, as happens in a fluidized product where the internal friction is low or zero. Where the stored contents slip on all the wall contact surfaces, the total mass of material is in motion and the pattern of behaviour is termed ‘Mass Flow’. Various other patterns are formed where the material slides on only part of the wall contact region. A symmetrical flow regime that has wall slip only on the region adjacent to the bin outlet is termed ‘Expanded Flow’. One that does not have wall slip until the flow channel has expanded within the static bed to meet the container walls at some height in the parallel section of the storage bin is called a 'Mixed Flow' pattern. 'Funnel Flow' is a two-stage flow regime that has a 'Core Flow' channel within the static bed supplied by material draining from the upper surface in an unconfined state at an inclination determined by its dynamic angle of repose. Non-symmetric flow channels take various forms, and are classed as 'Eccentric Flow'. This mode of behaviour introduces severe complexities into the understanding of solids flow and the boundary stress conditions. These forms are shown in Fig. 1. A simple variant is an 'Expanded Mixed Flow', when the flow channel of an 'Expanded Flow' form diverges to meet the container walls below the level of the stored contents.

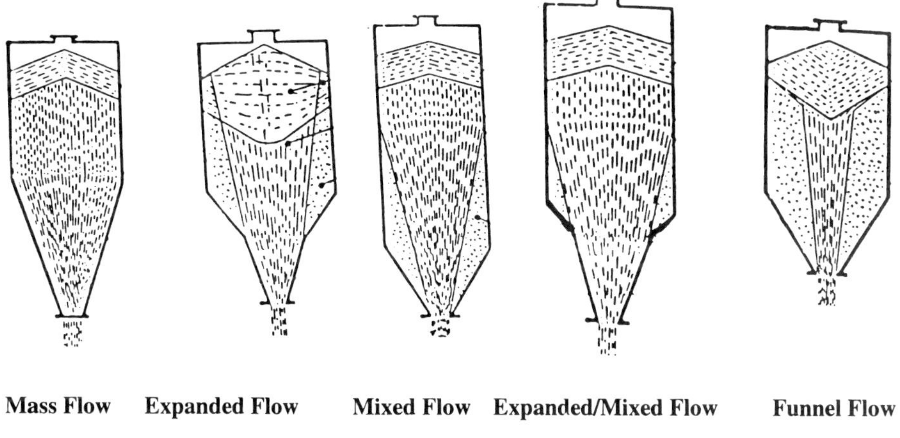

Fig 1. Forms of Flow Regimes from Storage Vessels

A further defining characteristic of a gravity flow channel is the manner in which the cross-section changes along the axis of flow. Flow in a parallel channel is termed 'Bed Flow'. The velocity profile across such a flow channel may not be uniform because of boundary friction, or because the extraction profile is dictated by a following section of the flow channel. Material invariably flows faster in the center of a converging flow channel than at the edges.

FORMS OF FLOW CONVERGENCE

The stresses acting upon the cross section of an element in a gravity flow channel are dependent upon the degree of convergence of the channel. The deformation imposed by a conical convergence causes a circumferential strain A/2 times the radial strain. Fig. 2.a. By comparison a plane flow channel offers only a confining boundary transverse to the converging strain, Fig. 2.b, a feature that explains why material will mass flow down shallower walls and through narrower outlets in a Vee shaped hopper than a cone. In greater contrast is an element that is not confined on one boundary surface. In this case the material can expand to allow deformation under considerably lower stresses. Fig. 2.c. The difference in structural nature between a 'dome' form of arch and a 'tunnel' arch allows a dome to be self supporting without a central region. This is why Cathedral domes can have a central hole, whilst a longitudinal arch, like a doorway, requires a 'keystone' to prevent the sides from falling in. The shear stress necessary to support material overhanging a circular hole is only half that needed to bear the weight of material overhanging a slot. For this reason the diameter of a hole has to be twice the width of a slot to cause a round arch to collapse in order for material to flow.

Although it is rarely practical to design an enclosed storage bin in a manner that allows unconfined flow throughout, the use of innovative geometry and inserts can provide a degree of relaxation of the flow channel confinement. This is achieved by creating walls that locally diverge slightly in one plane, and by fitting inserts that provides regions of reduced pressure on the boundary of the flow stream. This effect can be dramatically exploited in the crucial outlet region of a Vee shaped hopper, provided that the final outlet has fully 'live' flow across the total area. Even a small degree of transverse divergence in effective in allowing reliable mass flow at reduced wall angles and through smaller outlet widths.

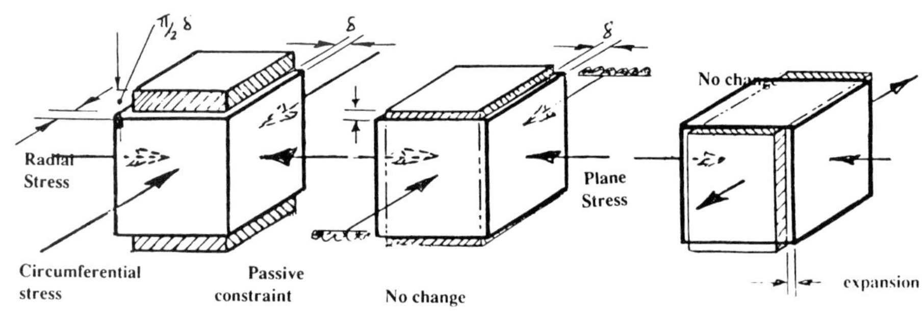

Conical Convergence **Planar Convergence** **Unconfined Convergence**

Fig. 2.a **Fig. 2.b** **Fig. 2.c**

Fig. 2 Principal Stress Systems on Flow Elements

The effectiveness of single plane convergence over conical convergence extends beyond advantages in wall inclination and outlet size. Whereas a conical flow channel can develop a stable 'rathole' by virtue of the hoop strength of material in the annulus, and deformation is necessary for the unconfined material to self clear, a Vee hopper can not sustain a full length 'rathole'. This is because the absence of material in the central region leaves the side portions unsupported, other than by frictional restraint on the inclined side walls. Once the central region has discharged, the residual material resting on the sides of a Vee hopper requires only that the walls are steeper than the angle of static friction for the material to slip to the outlet.

FLOW INSERTS

The design of flow inserts exploits the advantageous characteristics of these different failure stress methods. There are two main classes, those that alter the flow channel in a symmetrical manner and those which create regions of local pressure reductions to generate asymmetrical flow regimes. Symmetrical construction implies that the flow channel is to be regular in form and have an improved geometry for flow. This is done by reducing the rate of change of the horizontal stress field, or by changing it from Bi-axial to one that is uni-axial or has a reduced minor principal stress. An inverted Vee or a 'Vee and wall' insert, Figs 3 & 4, allows twin flow channels to be made with more gentle convergence than the original channel. Inverted cones and 'cones and cylinders', and the proprietary 'Easi-flow bin', Figs 5, 6 & 7, are examples of inserts that alter the form of flow from conical convergence to a pseudo planar form of flow channel.

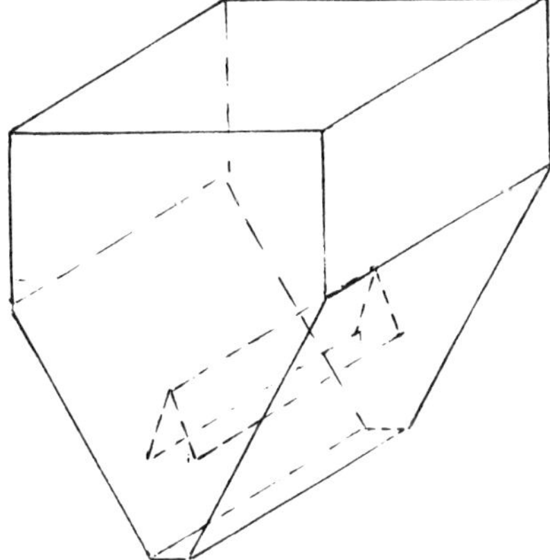

Fig. 4 Inverted Vee shaped insert

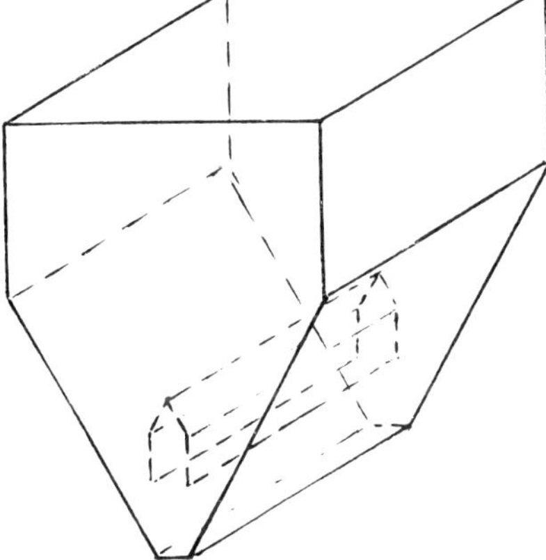

Fig. 5 Inverted Vee and body section

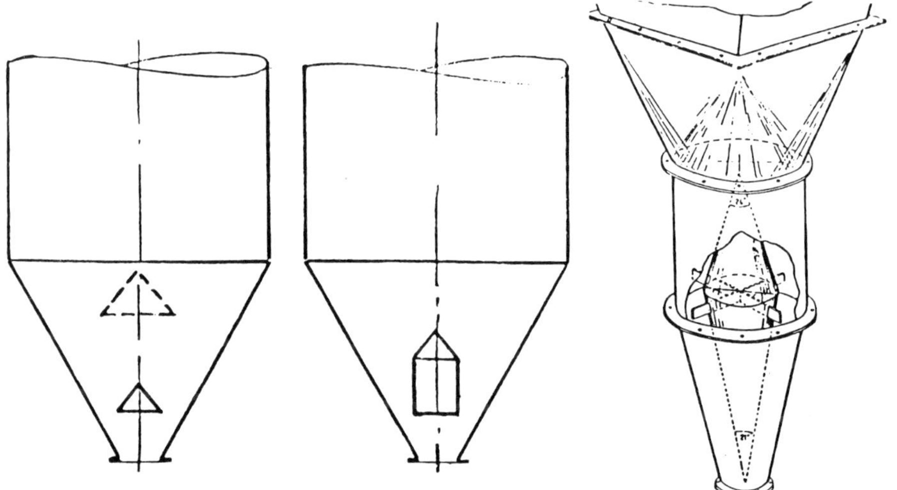

Fig. 5 Inverted Cone **Fig. 6 Inverted Cone and Body** **Fig. 7 'Easi-flow' Bin**

By contrast, the 'Binsert' or 'Cone-in-Cone' hopper type of insert is literally a hopper in a hopper. Fig. 8. The installation of a separate internal hopper of steeper form than the main vessel allows the central storage region to be served by a conical mass flow design and the outer region by a pseudo-Vee form of flow channel. Isolation of the 'core' pressure of the centrally flowing medium from the outer annulus of flow, enables mass flow deformation to take place in both channels within the original overall envelope of shape.

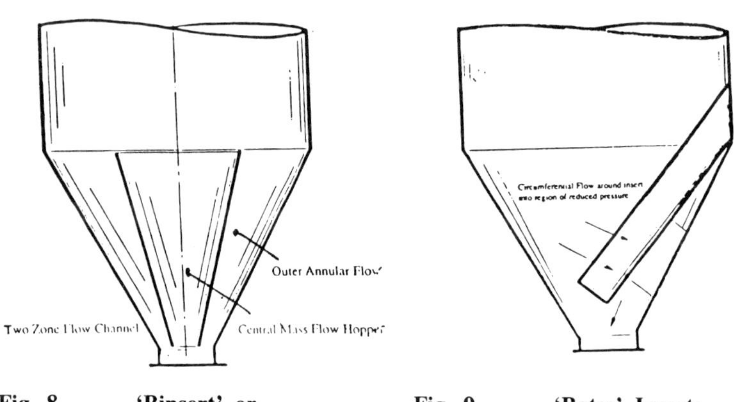

Fig. 8 'Binsert' or 'Cone-in-Cone' Insert **Fig. 9 'Bates' Inserts**

Flow inserts of non-continuous form are used to provide local regions of reduced pressure, into which the adjacent material can deform more readily. The is accomplished by re-directing the flow towards the outlet by obstructing the direct gravity favoured route. The 'Bates' inserts is mounted at an angle from the wall of a conical hopper to pass vertically over the outlet and at least part of the guide the flow channel towards the wall. Fig. 9.

The sector of reduced pressure in a portion of the circular cross section, behind a central member prevents direct cross flow, destroys the continuity of the hoop stress around the annulus of the cross section. As a consequence the remaining material closes towards the region of reduced pressure in a 'pincer-like' movement, giving a skewed form of flow as the material approaches the container outlet. The effect is progressively limited by the rapid increase in cross section of a cone in the direction above the outlet, otherwise this technique would allow mass flow at wall inclinations little above the angle of wall friction. In practice the pseudo wedge shape of flow channel and shielding of the outlet vicinity allows mass flow to develop local to the outlet at wall angles equivalent to Vee shaped flow channels.

An extension of this idea has been patented using multi-plate members. Fig. 10. Termed 'LynFlow' inserts, these fittings are wall mounted to shield part of the outlet and provide multiple peripheral regions of reduced flow pressures. The inner faces form a discontinuous hopper form but the total flow system is an integral pattern, rather than two or more channels.

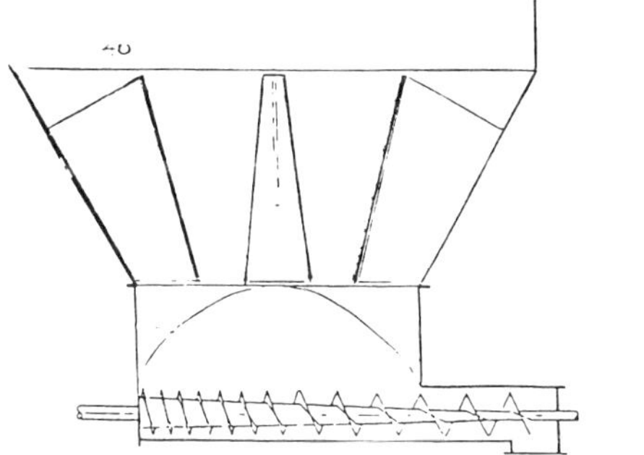

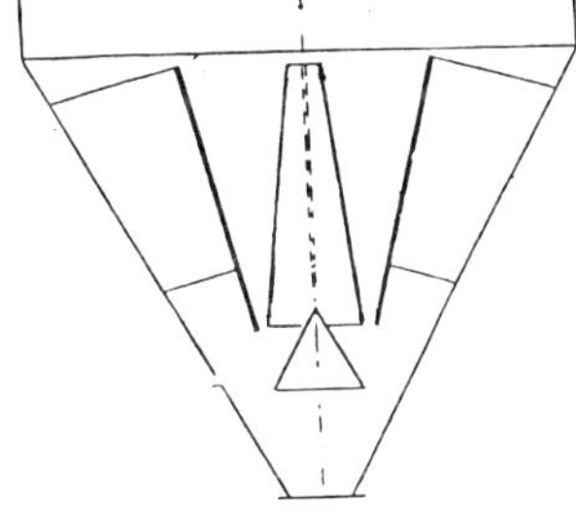

Fig. 10 'LynFlow' Inserts

OTHER INSERTS

Inserts are fitted for many reasons other than for flow. Some are of simple and obvious form, such as grids and deflectors to prevent wear and blockages by lumps. Other are more technically based, as with reverse sedimentation plates and rod frames for accelerated de-aeration of fluidized powders, Fig 11, and dispersing devices to avoid segregation effects from single point fill. Fig. 12 & 13, and dispersed extraction inserts to dilute segregation effects during the discharge of material from storage. Fig. 14

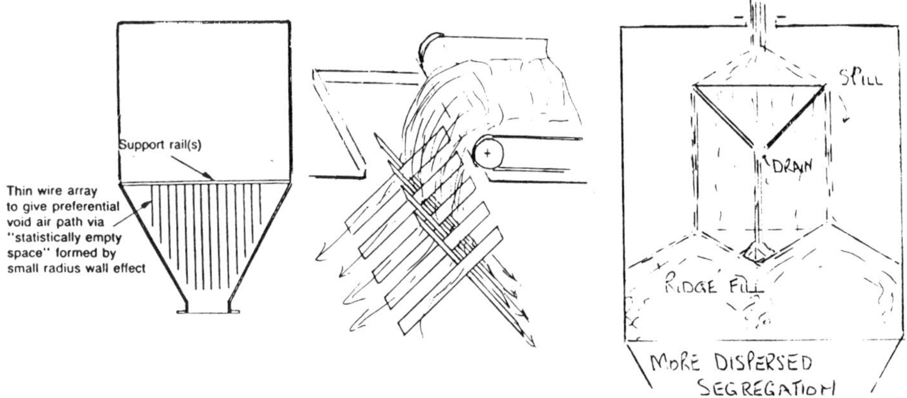

Fig 11 De-aeration Frame Fig. 12 Feed Dispersion Plates Fig. 13 In-feed Distributor

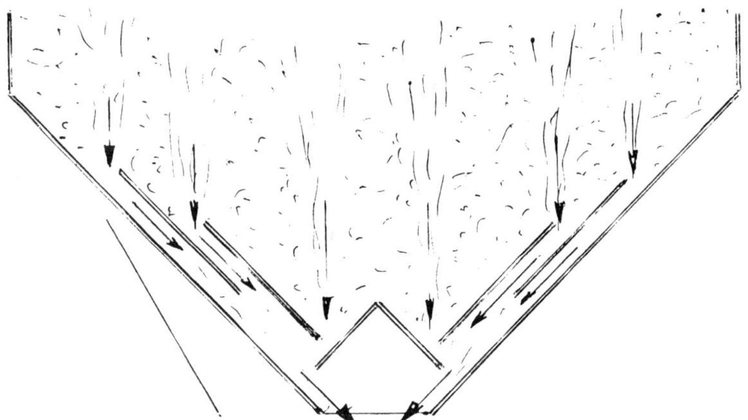

Fig. 14 Anti-Segregation Inserts for Multiple Point Extraction

SUMMARY

Gravity is reliable, universal and free. Brute force is non of these. Hopper problems are costly. Technical solutions are normally available. Right first time is best, but retrofits can be economical. Designers can exploit combinations of hopper geometry and insert systems to overcome solids handling problems and secure reliable performance at least costs. These must be based upon a clear understanding of the technology and measured relevant properties of the bulk material, particularly wall friction and shear strength.

S604/013/98

Proper design for reliable flow from hoppers and silos

H WRIGHT
Dr H Wright and Associates, Stockton-on-Tees, UK

Synopsis

In 1954, A.W. Jenike laid the foundation of a gravity flow theory of solids in bulk handling systems (1) and claimed solution to the problem of unreliable flow in bins, hoppers bunkers and silos, which, in the past, had been wholly remedied by the use of flow promotion devices. Examples are given of Jenike mass flow designs for coal storage that are working well after 20 years of continuous service. Over the last four decades, however, whilst this radical approach has become a proven practical reality around the world, application costs appear to mitigate its wider use.

Despite this, the Jenike shear cell test method is considered to be the "proper" route to take to ensure reliable flow from hoppers and silos and is the one to which all other shear cell design methodologies need to be correlated.

This paper concludes by investigating the possibility of analytical generation of key design parameters that would provide the same degree of hopper and silo success in the field, as the Jenike method. An initial appraisal of the authors Jenike flow property and mass flow geometric design database, of some 150 coals, indicates that whilst some evidence for improved correlation was found, in the area of ash percentage and lime/silica ratio, the search seems akin to that for the "holy grail".

1 INTRODUCTION

In the mid 1960's it was estimated that just to keep bins, hoppers, bunkers and silos flowing was costing the UK iron and steel industry about £1 million per year. Conservatively, this would scale up to at least £10 million today. In 1973, in a paper evaluating the Jenike design method, the author concluded that flow blockage was a major source of trouble (2). Despite the undoubted success of the Jenike approach, coupled with great improvements in the development and application of flow promotion equipment, in general, obtaining controlled flow from hopper and silo storage is considered to have become more difficult for industry to achieve over the years.

This paper begins by giving an overview of the author's wide range of success over the last 30 years, obtained by using Jenike's shear cell and flow property based geometric design concepts. These successful applications are but a small part of the world scene. Because application costs appear to mitigate the wider use of the Jenike method, the paper investigates the possibility of analytical generation of key Jenike related design parameters that may be able to give preliminary guidance on hopper and silo outlet sizes and wall slopes that will allow free flow.

2 THE JENIKE METHOD

The Jenike method to ensure reliable flow from hoppers and silos couples a flow theory for bulk solids to a practical technique for measuring their flow properties. The result of this is the prediction of the minimum acceptable values of the important hopper and silo design parameters, outlet size "D" and wall slope "Θ". Correct choice of these parameters ensures that any cohesive arch at the outlet just fails under its own weight.

Because Jenike's analytical method for instantaneous outlet size determination utilises the radial stress field, the result applies only to bulk solids in a dynamic stress state, i.e. that are flowing or have flowed. The basic Jenike method predicts the minimum outlet size for a moving material. This is a situation often overlooked or not known about which can lead to a flow blockage scenario following impact filling into an empty hopper or silo.

The Jenike method enables hoppers and silos to be designed for two types of flow:-

1. Mass flow - flow at the walls as well as in the central zone

2. Core (Funnel) flow - flow restricted to the central zone

The flow properties of a bulk solid are measured in a Flowfactor Tester which consists basically of a shear cell similar to that used in the study of soil mechanics. The split ring shear cell has an internal diameter of 95mm and a total depth of about 29mm.

The originally recommended sample size of bulk solid for use in the shear cell (3) was that fraction passing 20 mesh US Sieve Series (- 841 micron). The size fraction was arbitrarily chosen by Jenike on the grounds that it is the fine particles that produce the strength of a cohesive arch. If the bulk solid is wet, the moisture content at which the shear tests should be carried out can be chosen on any of the following three bases:

(a) The actual moisture content occurring in practice, if constant.

(b) The moisture content giving the bulk solid its greatest shear strength, if variable.

(c) The maximum moisture content with which the plant is able to cope.

Although the Jenike shear cell is probably the most widely used flow property test method in the world today, it does suffer from a number of disadvantages as follows:-

(i) The limited displacement necessitates laborious sample preparation prior to testing,

(ii) The unequal stress distribution gives progressive failure,

(iii) The sample is forced to shear on a predetermined plane which is not necessarily the weakest one,

(iv) The method becomes inaccurate at low loads due to tilting of the lid. These low loads operate in the lower sections of the bunker and often extrapolation of test data is required for outlet diameters < 0.5m.

A major disadvantage considered by the author is the original Jenike preconsolidation technique. This consists of twisting the sample a predetermined number of times and then subjecting it to shear until a value of 95% of that required to cause failure is reached. In this way the sample is taken as near to the failure condition as possible without actually causing failure. An individual yield locus is determined by re-shearing preconsolidated samples at lower values of the normal load.

Because of difficulty experienced in trying to apply the Jenike sample preparation technique a new method of sample preparation became necessary. An alternative, called the compaction overload method, is suggested in a paper by Williams and Birks (4) and was utilised by the author in his PhD thesis evaluating the Jenike design method (5). In essence, for each yield locus, this enables the determination of the precise ratio of consolidating load to maximum normal load that will generate a stress/strain curve that is neither under nor overconsolidated and which has minimum volume change (i.e. a yield locus end point). Armed with this data the designated yield loci are constructed by shearing to failure, samples preconsolidated to the relevant compaction overload ratio under a range of normal loads starting from the maximum and reducing down to the lowest value required.

3 THREE DECADES OF SUCCESSFUL MASS FLOW DESIGNS

Since 1968, the author, using the modified Jenike shear cell testing procedure and design methodology indicated above, has determined the flow properties of over 500 bulk solids and produced mass flow designs ranging from a 1000 m^3 silo for arsenic trioxide to 4 x 3000t silos for fine zinc ore. This flow property testing has resulted in the geometric design of hundreds of mass flow bins, hoppers, bunkers and silos valued at many millions of pounds sterling (£).

In 1978 (6), the author outlined details of 88 mass flow designs for 9 major capital developments. These ranged in capacity from 76t - 8170t storing some 70 steelworks bulk solids including 28 iron ores and 19 coals/coal blends. The designs were built and commissioned over the period 1969 - 1974. It is believed that the majority of these are still working well. Up to the present over 150 coals from some sixteen countries have also been tested.

3.1 Coal storage - the past

Over the years, the design of mass flow bins, hoppers, bunkers and silos for wet coal has provided the Jenike method with a "second to none" strongly practical success base. In 1973 the author was asked to assist in the design of 18 x 1000t capacity coal blending silos for British Steel's Dawes Lane coke oven complex at Scunthorpe. Flow properties were measured for nine UK coals at critical (most cohesive) surface moisture contents ranging from 14 - 20% (wet weight). The mass flow geometric design data for the nine coals tested, published in 1981 (7), is given below.

Design Parameters for the Range of 9 Coals Tested	Maximum	Minimum
Bulk Density (kg/m^3)	890	712
Instantaneous Outlet Diameter (m)	1.22	0.84
Time/Impact Filling Outlet Diameter (m)	1.73	1.32
Wall Slope to the Horizontal - mild steel	79°	67°
Wall Slope to the Horizontal - cast glass	70°	59°
Wall Slope to the Horizontal - granofinish concrete	75°	65°

The mass flow design chosen for the 18 concrete silos incorporated (i) a 9.144m diameter shell, (ii) a cone angle of 70° to the horizontal, (iii) a cast glass lining in the cone and (iv) a 1.73m diameter outlet. The total silo height was 30m which included a 10m deep cone section. It is reported that these silos continue to work well even with a different range of coals from those originally specified.

3.2 Coal storage - the present

In 1992, the conceptual geometric design for a 2,600t capacity twin outlet mass flow steel coal silo, built for National Power plc at Bristol, was verified using average coal flow property data for 27 UK and imported steam coals. It is also reported that this silo continues to work well even with a different range of coals from those used in the verification analysis.

4 FLOW PROPERTY RELATED ANALYTICAL DESIGN OF HOPPERS AND SILOS

4.1 Introduction

The authors large database of Jenike shear cell flow properties and associated mass flow designs for over 150 coals allows comparison between:-

(a) the key flow parameter of instantaneous mass flow outlet size "D",

(b) the key flow property directly related to mass flow wall slope which is the critical value of effective angle of internal friction (δ), and the analytical variables (i) bulk density, (ii) free moisture content, (iii) mean particle size, (iv) % of sieve size passing 2mm, (v) % ash, and (vi) % lime/silica ratio.

4.2 Results

Figure 1 shows a graph of bulk density v percentage variation of instantaneous mass flow outlet size "D" about its mean (ie 100%) for 147 UK and imported coals. The scatter of the data is immense as indicated by the extremely low linear regression correlation coefficient of 0.367. This data clearly shows that there is no relationship between bulk density and the Jenike instantaneous mass flow outlet diameter.

Figure 2 shows a graph of bulk density v percentage variation of critical value of effective angle of internal friction "δ" about its mean (ie 100%) for 142 UK and imported coals. The scatter of the data is poor as indicated by the very low linear regression correlation coefficient of 0.541. This data shows that there is little or no relationship between bulk density and the Jenike effective angle of internal friction and hence mass flow wall slope.

Figure 3 shows a graph of free moisture v percentage variation of instantaneous mass flow outlet size "D" about its mean (ie 100%) for 146 UK and imported coals. The scatter of the data is again immense as indicated by the extremely low linear regression correlation coefficient of 0.289. This data clearly shows that there is no relationship between free moisture content and the Jenike instantaneous mass flow outlet diameter.

Figure 4 shows a graph of free moisture v percentage variation of critical value of effective angle of internal friction "δ" about its mean (ie 100%) for 141 UK and imported coals. The scatter of the data is again immense as indicated by the very low linear regression correlation coefficient of 0.404. This data clearly shows that there is no relationship between free moisture content and the Jenike effective angle of internal friction and hence mass flow wall slope.

Figure 5 shows a graph of mean size v percentage variation of instantaneous mass flow outlet size "D" about its mean (ie 100%) for 6 UK coals. The scatter of the data is immense as indicated by the extremely low linear regression correlation coefficient of 0.0609. This data clearly shows that there is no relationship between mean size and the Jenike instantaneous mass flow outlet diameter.

Figure 6 shows a graph of mean size v percentage variation of critical value of effective angle of internal friction "δ" about its mean (ie 100%) for 6 UK coals. The scatter of the data is again immense as indicated by the very low linear regression correlation coefficient of 0.3432 This data clearly shows that there is no relationship between mean particle size and the Jenike effective angle of internal friction and hence mass flow wall slope.

Figure 7 shows a graph of percentage passing 2mm sieve size v percentage variation of instantaneous mass flow outlet size "D" about its mean (ie 100%) for 6 UK coals. The scatter of the data is again immense as indicated by the extremely low linear regression correlation coefficient of 0.0412. This data clearly shows that there is no relationship between percentage passing 2mm sieve size and the Jenike instantaneous mass flow outlet diameter.

Figure 8 shows a graph of percentage passing 2mm sieve size v percentage variation of critical value of effective angle of internal friction "δ" about its mean (ie 100%) for 6 UK coals. The scatter of the data is again immense as indicated by the very low linear regression correlation coefficient of 0.352. This data clearly shows that there is no relationship between percentage passing 2mm sieve size and the Jenike effective angle of internal friction and hence mass flow wall slope.

Figure 9 shows a graph of percentage ash v percentage variation of instantaneous mass flow outlet size "D" about its mean (ie 100%) for 4 overseas coals. The scatter of the data is poor as indicated by the low linear regression correlation coefficient of 0.578. This data shows that there is little or no relationship between percentage ash and the Jenike instantaneous mass flow outlet diameter.

Figure 10 shows a graph of percentage ash v percentage variation of critical value of effective angle of internal friction "δ" about its mean (ie 100%) for 4 overseas coals. The scatter of the data is wide as indicated by the fairly low linear regression correlation coefficient of 0.771 This data shows that there is little relationship between percentage ash and the Jenike effective angle of internal friction and hence mass flow wall slope.

Figure 11 shows a graph of percentage lime/silica v percentage variation of instantaneous mass flow outlet size "D" about its mean (ie 100%) for 4 overseas coals. The scatter of the data is again quite wide as indicated by the very low linear regression correlation coefficient of 0.376. This data clearly shows that there is no relationship between percentage lime/silica and the Jenike instantaneous mass flow outlet diameter.

Figure 12 shows a graph of percentage lime/silica v percentage variation of critical value of effective angle of internal friction "δ" about its mean (ie 100%) for 4 overseas coals. The scatter of the data is fairly wide as indicated by the quite low linear regression correlation coefficient of 0.805 This data shows that there is little relationship between percentage lime/silica and the Jenike effective angle of internal friction and hence mass flow wall slope.

5 CONCLUSIONS

1 Over the last three decades around the world, the Jenike shear cell flow property test method for bins, hoppers, bunkers and silos has produced many geometric designs having a high level of efficiency. These have regularly produced 100% useful capacity without arching, ratholing or the need for flow promotion equipment of any kind.

2 Since 1968, the author has determined the flow properties of over 500 bulk solids and produced mass flow designs for bulk solids ranging from arsenic trioxide to zinc ore. The flow property testing has resulted in the generation of cost effective geometric design of hundreds of mass flow bins, hoppers, bunkers and silos, some of which have successfully handled over 20 million tonnes of stored product with very few problems.

3 The Jenike shear cell test method is considered to be the "proper" route to take to ensure reliable flow from hoppers and silos.

4 Other shear cell equipment on the market needs to produce test data that correlates well with Jenike if similar practical success in the field is desired.

5 Regression analysis of a wide range of flow property data for coal has indicated that there is no meaningful correlation between the key Jenike design parameters for mass flow and analytical variables such as bulk density and free moisture content.

6 Whilst the author believes that the use of analytical means of any sort to arrive at a viable

mass flow design is akin to searching for the "holy grail", there does appear to be some improvement in correlation coefficients for coal in the area of percentage ash and lime/silica ratio.

REFERENCES

1 Jenike, A W, Flow of Solids in Bulk Handling Systems, Bulletin 64 of the Utah Engineering Experiment Station, University of Utah, USA, 1954.

2. Wright, H, An evaluation of the Jenike bunker design method, Iron & Steel International, June 1973, pp 252 - 258.

3 Jenike, A W, Storage and Flow of Solids, Bulletin 123 of the Utah Engineering Experiment Station, University of Utah, USA, 1964.

4 Williams, J C and Birks, A H, The Preparation of Powder Specimens for Shear Cell Testing, Rheologica Acta, Band 4, Heft 3, 1965.

5. Wright, H, Bunker Design for Iron Ores, PhD Thesis, Dept of Chemical Engineering, University of Bradford, UK, 1970.

6. Wright, H, BSC's contribution to the design and operation of mass flow bunkers, Iron & Steel International, August 1978, pp 233 - 238.

7. Wright, H, Scunthorpe coal silos show off BSC's design skills, Iron & Steel International, June 1981, pp143 - 146.

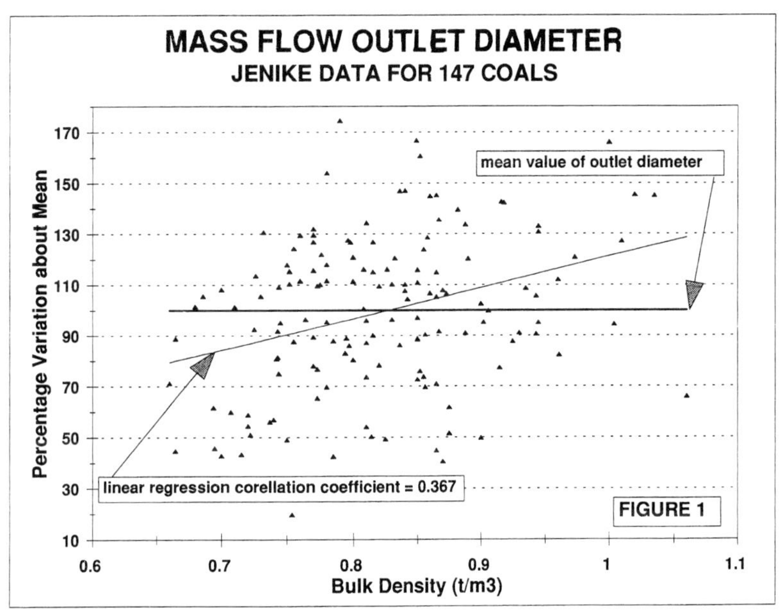
MASS FLOW OUTLET DIAMETER
JENIKE DATA FOR 147 COALS
mean value of outlet diameter
linear regression corellation coefficient = 0.367
FIGURE 1
Percentage Variation about Mean
170
150
130
110
90
70
50
30
10
0.6
0.7
0.8
0.9
1
1.1
Bulk Density (t/m3)

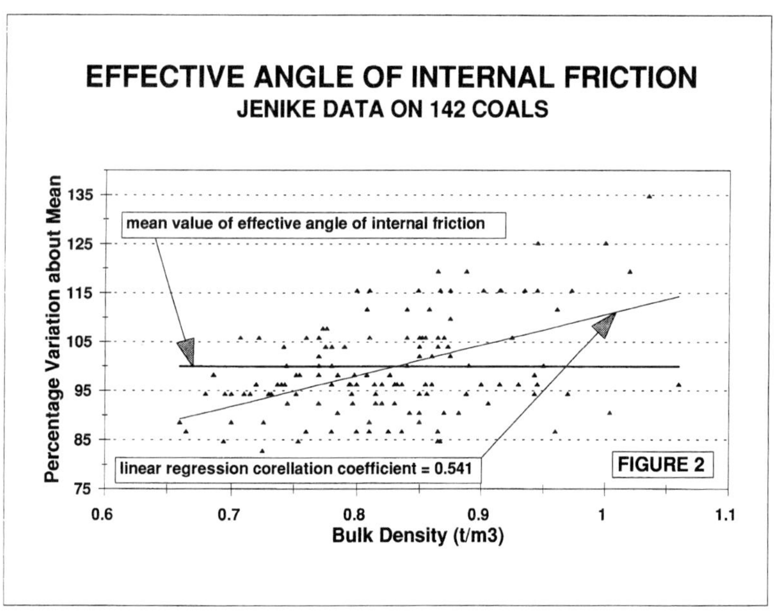
EFFECTIVE ANGLE OF INTERNAL FRICTION
JENIKE DATA ON 142 COALS
mean value of effective angle of internal friction
linear regression corellation coefficient = 0.541
FIGURE 2
Percentage Variation about Mean
135
125
115
105
95
85
75
0.6
0.7
0.8
0.9
1
1.1
Bulk Density (t/m3)

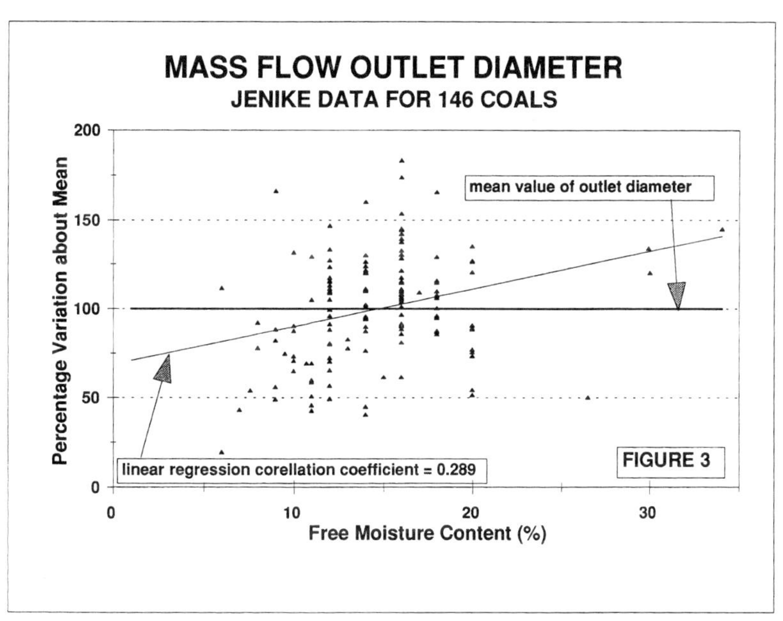
MASS FLOW OUTLET DIAMETER
JENIKE DATA FOR 146 COALS
mean value of outlet diameter
linear regression corellation coefficient = 0.289
FIGURE 3
Percentage Variation about Mean
Free Moisture Content (%)
200
150
100
50
0
0
10
20
30

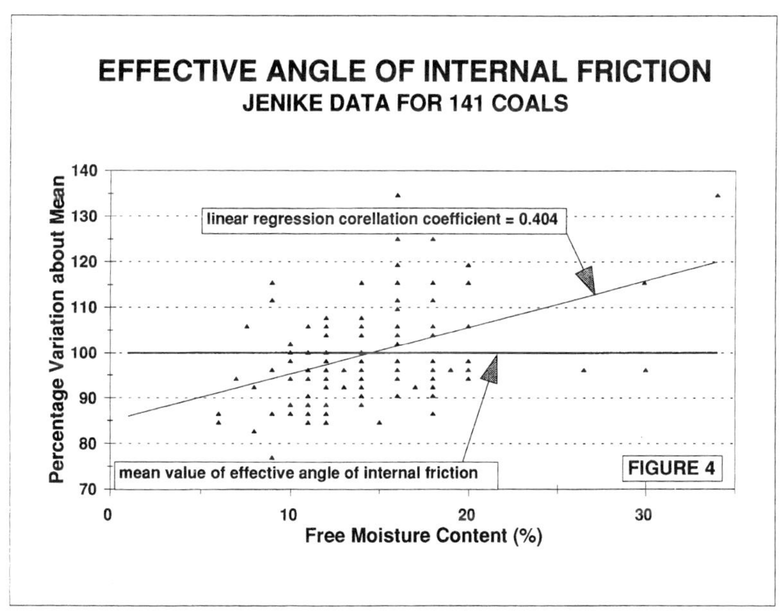
EFFECTIVE ANGLE OF INTERNAL FRICTION
JENIKE DATA FOR 141 COALS
linear regression corellation coefficient = 0.404
mean value of effective angle of internal friction
FIGURE 4
Percentage Variation about Mean
Free Moisture Content (%)
140
130
120
110
100
90
80
70
0
10
20
30

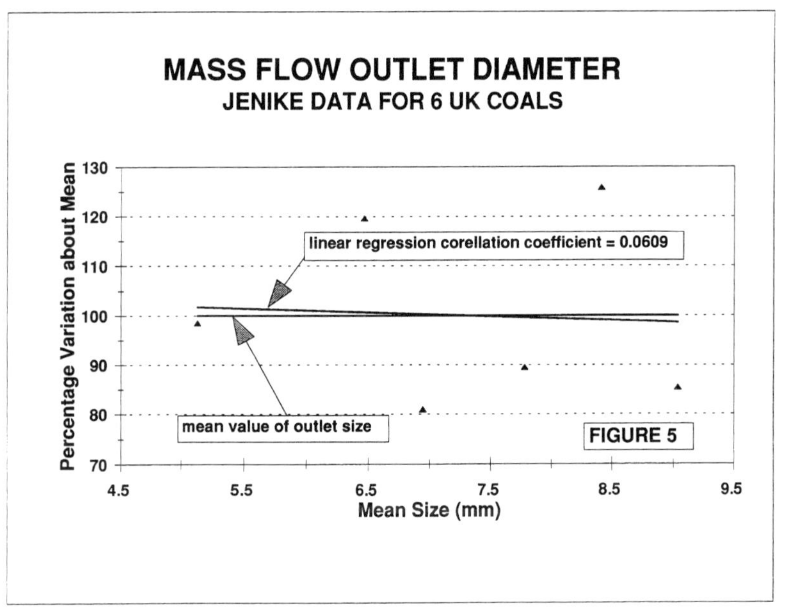
MASS FLOW OUTLET DIAMETER
JENIKE DATA FOR 6 UK COALS
Percentage Variation about Mean
130
120
110
100
90
80
70
linear regression corellation coefficient = 0.0609
mean value of outlet size
FIGURE 5
4.5
5.5
6.5
7.5
8.5
9.5
Mean Size (mm)

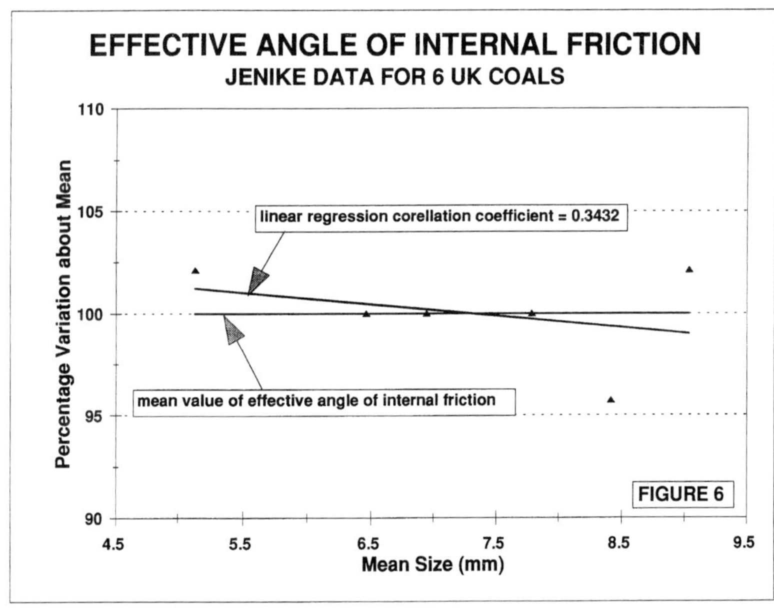
EFFECTIVE ANGLE OF INTERNAL FRICTION
JENIKE DATA FOR 6 UK COALS
Percentage Variation about Mean
110
105
100
95
90
linear regression corellation coefficient = 0.3432
mean value of effective angle of internal friction
FIGURE 6
4.5
5.5
6.5
7.5
8.5
9.5
Mean Size (mm)

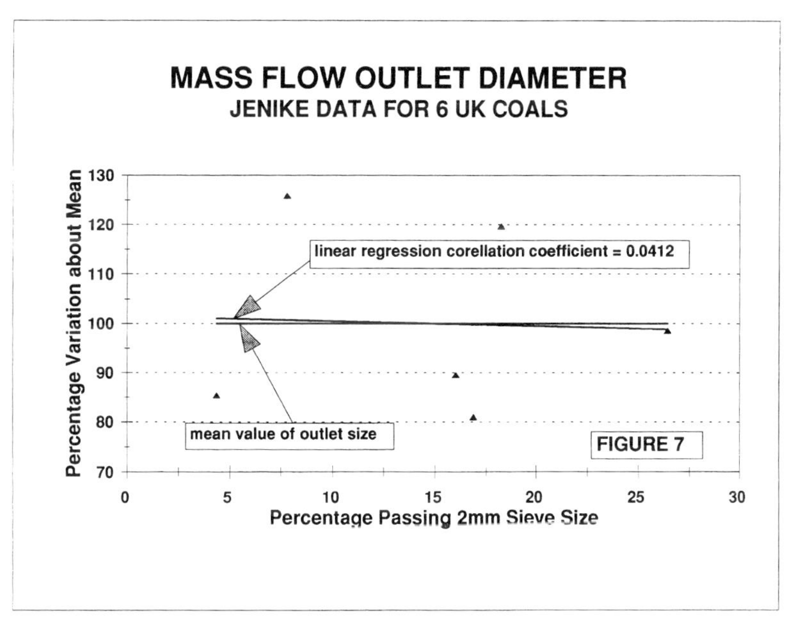
MASS FLOW OUTLET DIAMETER
JENIKE DATA FOR 6 UK COALS
Percentage Variation about Mean
130
120
110
100
90
80
70
linear regression corellation coefficient = 0.0412
mean value of outlet size
FIGURE 7
0
5
10
15
20
25
30
Percentage Passing 2mm Sieve Size

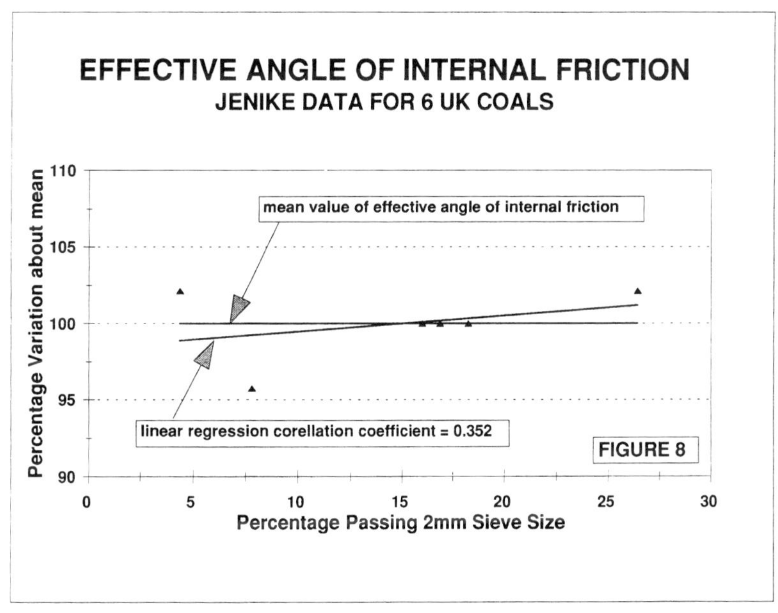
EFFECTIVE ANGLE OF INTERNAL FRICTION
JENIKE DATA FOR 6 UK COALS
Percentage Variation about mean
110
105
100
95
90
mean value of effective angle of internal friction
linear regression corellation coefficient = 0.352
FIGURE 8
0
5
10
15
20
25
30
Percentage Passing 2mm Sieve Size

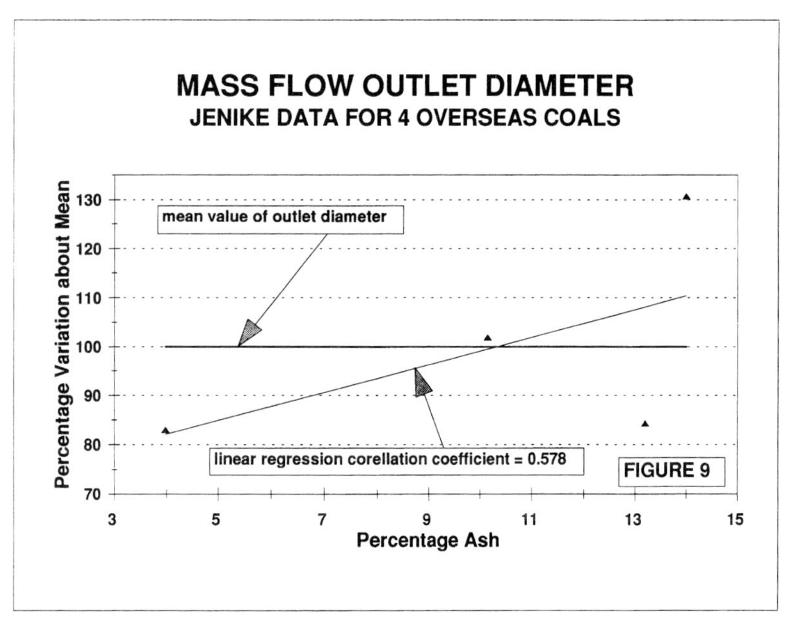
MASS FLOW OUTLET DIAMETER
JENIKE DATA FOR 4 OVERSEAS COALS
Percentage Variation about Mean
130
120
110
100
90
80
70
mean value of outlet diameter
linear regression corellation coefficient = 0.578
FIGURE 9
3
5
7
9
11
13
15
Percentage Ash

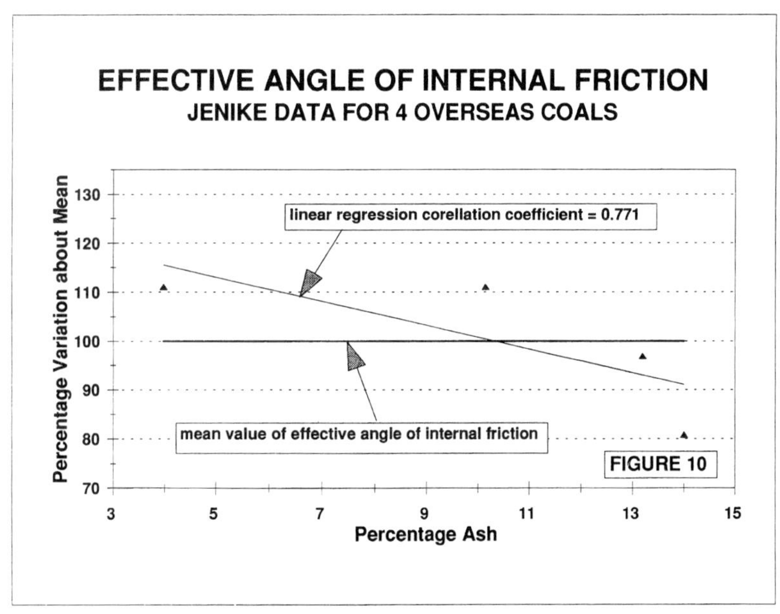
EFFECTIVE ANGLE OF INTERNAL FRICTION
JENIKE DATA FOR 4 OVERSEAS COALS
Percentage Variation about Mean
130
120
110
100
90
80
70
linear regression corellation coefficient = 0.771
mean value of effective angle of internal friction
FIGURE 10
3
5
7
9
11
13
15
Percentage Ash

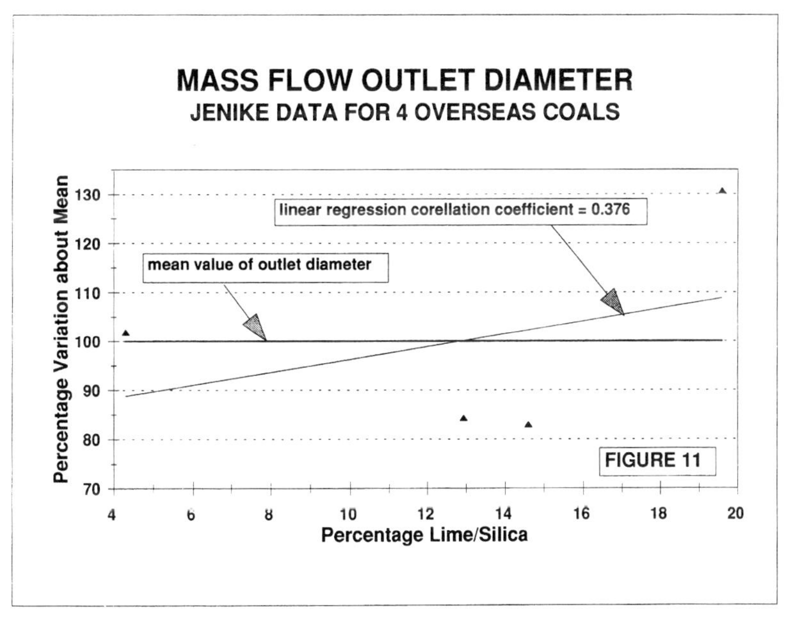
MASS FLOW OUTLET DIAMETER
JENIKE DATA FOR 4 OVERSEAS COALS
Percentage Variation about Mean
130
120
110
100
90
80
70
linear regression corellation coefficient = 0.376
mean value of outlet diameter
FIGURE 11
4
6
8
10
12
14
16
18
20
Percentage Lime/Silica

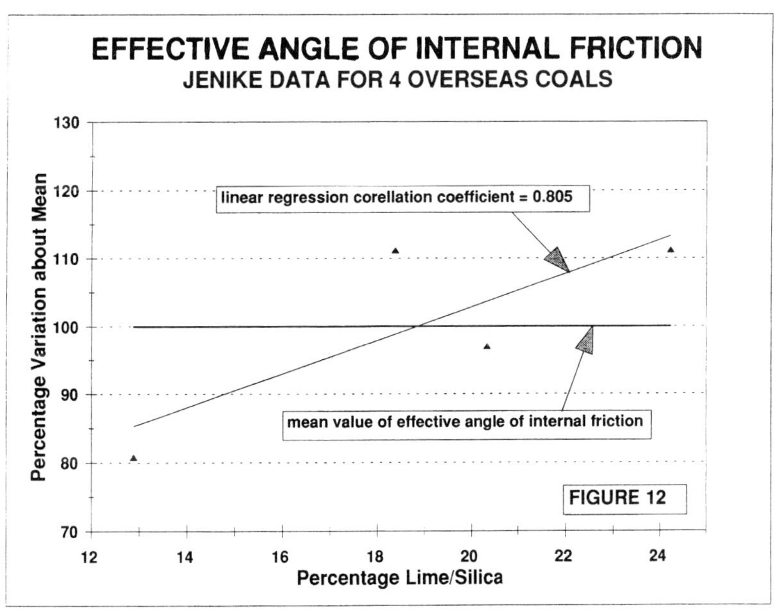
EFFECTIVE ANGLE OF INTERNAL FRICTION
JENIKE DATA FOR 4 OVERSEAS COALS
Percentage Variation about Mean
130
120
110
100
90
80
70
linear regression corellation coefficient = 0.805
mean value of effective angle of internal friction
FIGURE 12
12
14
16
18
20
22
24
Percentage Lime/Silica

S604/014/98

Strategy for selecting the solution

M BRADLEY
University of Greenwich, London, UK

SYNOPSIS

It is common to have to deal with existing hoppers and silos, which do not discharge reliably. The engineer faced with such a problem needs some sort of understanding of how to select the most appropriate of the many possible solutions available. The choice depends upon consideration of:-

- The material and its flow behaviour,
- The process requirements,
- The size of the installation,
- The capital available, and
- The prospect for future utilisation of the equipment.

The many types of mechanical extractors, vibrating flow aids, pneumatic devices and other options each have a place, but the selection of the wrong approach can lead to disappointment or even a worsening of the problem. However, even given the choice of the appropriate device, the way in which it is applied can make the difference between success and failure.

This paper seeks to give a general understanding of the principle of employing aids to flow, and develop some guidance to the engineer as to how to approach the problem, choose the solution most likely to be successful, and apply it in the best way.

1. BASIC PRINCIPLES

1.1 Core flow discharge and hopper obstructions

There are basically two possible flow patterns in hoppers, i.e. core flow and mass flow, as depicted below. Although mass flow has some useful advantages, the vast majority of vessels discharge in core flow. This is because achieving mass flow requires some very careful design, usually following flow testing of the product being stored; in practice this is not often done, consequently the "default" pattern of core flow is instead arrived at. The specific advantages of mass flow will be explored later in this paper.

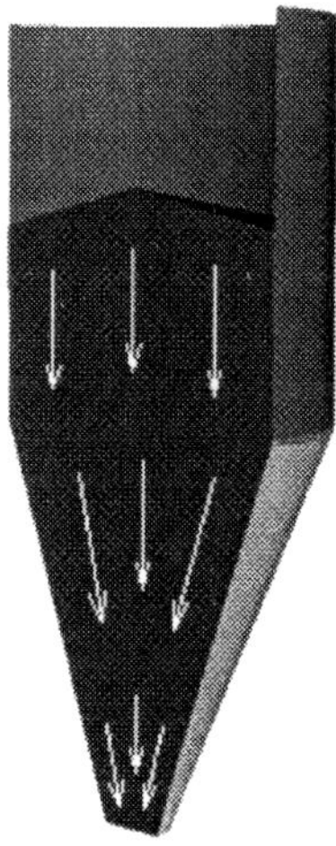

Fig. 1a Mass flow

Material all in motion, sliding on the cone of the vessel and coming out in order of input

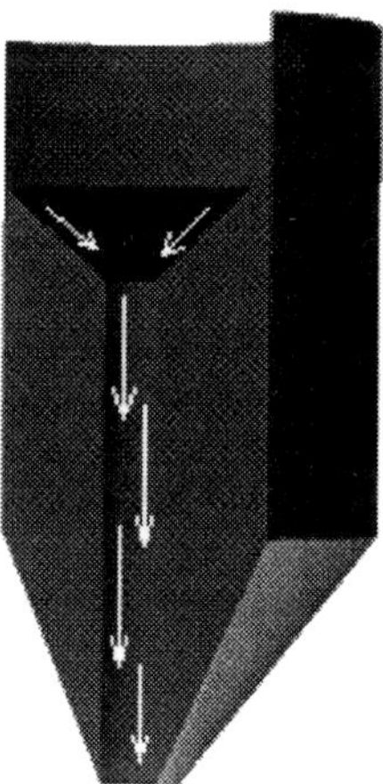

Fig. 1b Core flow

Material sloughs off the top surface down an angle of repose and flows down through a central flow channel

1.2 Ensuring reliable core flow

Failures to discharge from core flow hoppers can occur as a result of two different types of blockage, the stable arch or the rat-hole as shown below.

Fig. 2a Stable arch

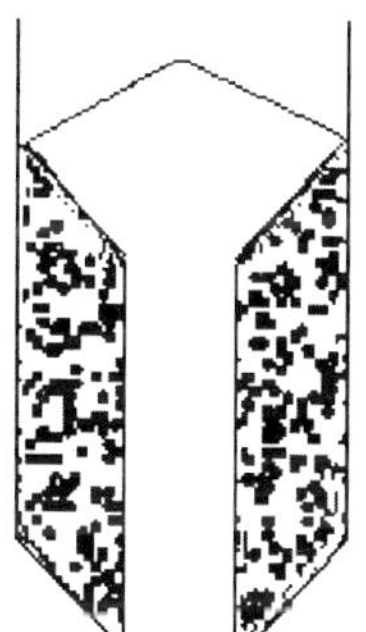

Fig. 2b Stable rat-hole

These are both stable structures in which the strength of the bulk solid is higher than the stresses in the structure (from its self-weight), meaning that the structure will not collapse under its own weight. The stresses in the structures increase as the size of the structures (arch or rat-hole) increase; the size of the structure is dictated by the outlet size of the vessel, hence it follows that if the outlet size of the vessel is continually increased, there will eventually arise a situation where the stresses in the structure overcome the strength of the material and the structure collapses, allowing flow to occur.

Any given commodity has a certain strength characteristic, which effectively gives it a maximum dimension across which it can arch, and a maximum size of rat-hole which it can support. These dimensions depend to a degree on the size of the bin and hence the pressure applied to the material at the bottom to compact it, and also on the maximum time in residence between complete emptyings-out of the bin (this "time consolidation" effect can also strengthen the material).

The important point to understand from the above is this: *For a given commodity in a given vessel, there is a certain maximum arching dimension and a certain maximum rat-hole dimension which the commodity can support.* As it happens, the maximum rat-hole dimension is usually above the maximum arching dimension. If the outlet of the vessel is below the arching dimension of the material, an arch will result when material is withdrawn; if the outlet of the vessel is above the arching dimension, but below the rat-hole dimension then a stable rat-hole will form once the material in the centre of the bulk has flowed out.

For a core flow bin to discharge successfully by gravity, without flow aids, the outlet size would need to be above the maximum stable rat-hole dimension of the commodity being stored. In bins where there is not problem with flow, this is the case. However where there is a flow problem, clearly it is not.

One option is to enlarge the outlet dimension above the maximum stable rat-hole dimension. If this is too big to connect directly to the equipment under the vessel, then a mechanical feeder of some kind is needed to take the commodity from this large hopper outlet and feed it into the smaller equipment underneath. This approach, whilst workable, often is not possible or desirable, especially as the stable rat-hole dimension of the commodity can often be several feet with cohesive materials.

1.3 Successful application of flow aids

The objective of a flow aid is effectively to turn the lower section of the hopper into a feeder, by artificially encouraging flow of the contents where such flow would not occur under gravity alone. *An effective flow aid must therefore encourage flow down to the mouth of the hopper, and all the way up to a diameter above the stable rat-hole dimension of the commodity.*

One point to note is that if the commodity has a high tendency to "time consolidate" (i.e. gain strength with time) then the maximum stable rat-hole diameter will be very large, not uncommonly as large as the vessel containing it. This is a particular problem with a core flow hopper, since with that flow pattern, the material up around the hip will remain static all the time until the hopper is actually right emptied out. In such a case, for a flow aid to be successful, it has to activate the entire height of the converging section. However, for materials with such a high tendency to time consolidate, it is probably more economic in the long term to change to a mass flow hopper for reasons which will become apparent later.

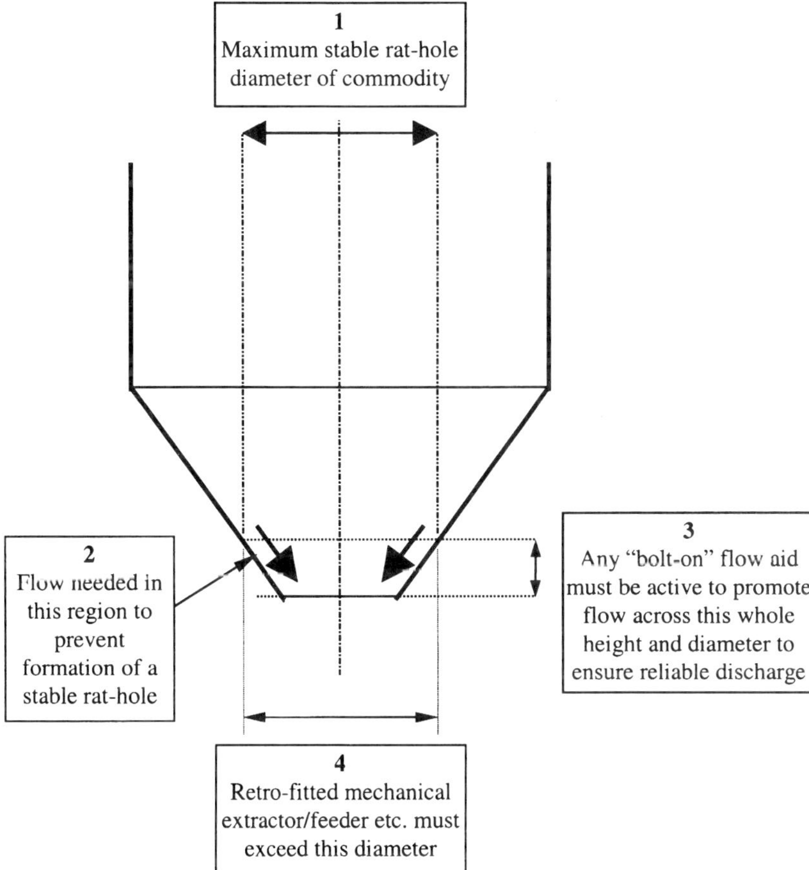

Fig. 3 Application of flow aids to a core flow hopper for successful discharge

The many different types of flow aid work in different ways to achieve the same objective. Their ranges of applicability, and particular points to be watched when using them, will be explored below. Of course, the critical rat-hole dimension of the material in the hopper is often not known (although it can be determined by flow property measurement) but a general understanding of the above principle is necessary to ensure the successful use of discharge aids.

1.4 Available options

Briefly, the most commonly employed aids to improved flow are:-

Vibrating discharge aids (internal and external)
Aeration devices
Mechanical extractors
Mass flow
Inserts and linings

Altering the product

These will be taken in turn and their applicability examined.

2. VIBRATING DISCHARGE AIDS

Vibration will be explored in some detail since it is probably the most commonly-applied aid to discharge. It works extremely well to control the flow of free-flowing materials; ironically, discharge aids are rarely needed for such materials! However, they can be used for quite a wide variety of commodities provided the commodities are not either:-

(a) Highly cohesive, such that compressing them leads to a big gain in strength; Such materials like wet or damp sludges, or anything which forms a strong "snowball" when pressed in the hand, tend to be compacted by vibration; so applying vibration to such materials can frequently make the problem worse;

(b) Highly elastic, so that they just absorb the vibration instead of transmitting it through the bulk, such as bran, germ or wood chips; or

(c) Very fine (say 30 microns or less) which have a tendency to be very variable in their characteristics depending upon what state of settlement or aeration they are in; examples include cement, titanium dioxide or powdered carbon black.

2.1 External vibrators

Vibrators bolted to the outside of a vessel vibrate the wall, thereby reducing the friction between hopper and contents and (to a small degree) reducing the strength of the bulk. The reduced wall friction undermines any arch and puts an inwards pressure on any rat-hole, helping to collapse it.

To be effective, external vibrators must be applied in the right place, that is to vibrate the hopper wall in the region between the outlet and the critical rat-hole diameter, as shown in fig. 3 above. This presents something of a problem, since the area around the mouth of the hopper is its strongest part; frequently, it is bolted to a rotary valve or other heavy piece of equipment. Consequently, getting the vibration to transmit into this part of the structure is not easy. Nevertheless it is important that the vibration is carried as far down the wall towards the outlet as possible, as this is the most critical area for flow (being the smallest). If vibrators are fitted too high up, then flow will be encouraged high up which will simply serve to apply pressure to the material lower down near the outlet of the hopper and consolidate it, making it harder to discharge.

In general, it is best to start with vibrators low down on the hopper since the vibration will transmit upwards into the more flexible part of the structure, more readily than downwards.

Such external vibrators are usually installed as a first retro-fit option to a troublesome hopper; they are quite cheap, readily available and easy to install. Positioning has been mentioned; sizing is rather more hit-and-miss but suppliers can advise. Experience has shown that they are effective at overcoming slight flow problems (where discharge is mostly adequate but stops now and again and can be restarted with a bit of hammering). It can deal with more regular problems as long as the material is not highly cohesive. Often they do not need continuous running; just a fifteen-second period when flow is first initiated, or perhaps the same every minute during discharge. It is always better to start with too little vibration and increase it if necessary, to avoid the danger of compacting the material especially if the flow rate is limited under the hopper by a rotary valve or feeder.

Vibrators must always be sequenced so that they only come on when the material is being taken away from the hopper outlet; applying vibration to any material when it is confined will simply consolidate it.

Given the cheapness and ease of installation, the application of external vibrators is always worth a try unless the material is known to be highly cohesive or elastic or the discharge problem is extremely severe.

Some drawbacks are:-

- They can introduce quite substantial noise into the workplace;
- If oversized or badly installed, they can damage the hopper by metal fatigue.
- They are not good for overcoming severe problems of hopper blockage or rat-holing; trying to use a high level of vibration by this means will lead to damage to the hopper and often serve to settle and compact and long-term resident material in the hopper.

2.2 Internal vibrators

Where more severe flow problems occur it is better to install a well-designed internal vibrating flow aid

2.2.1 Vibrating cones

In one realisation of these devices, the internal cone contains an air-driven vibrator and mounts inside the existing hopper cone on flexible mounts; examples include the well-known “Soliflo” or “Matcon” devices. In other versions (e.g. the “bin activator”) the cone is mounted rigidly inside a conical or dish-shaped replacement hopper bottom, itself mounted on flexible mounts and driven by external vibrator motors. The first type can sometimes be retro-fitted to an existing bin, whereas the second type requires the bottom of the hopper to be removed. Either can be designed in to new installation quite effectively.

In either case, the internal cone vibrates much more effectively than a hopper wall with a bolt-on vibrator. This promotes flow of the material right across the hopper from wall to wall, adjacent to thc cone, by allowing flow down the surface of the cone and making it impossible for the material to arch between the hopper wall and the moving cone. Provided the diameter of the vibrating cone is nearly as large as the rat-holing dimension of the material, this will promote flow very well.

Vibration is applied continuously or intermittently all the time discharge is required, and when vibration is turned off the material arches between the static cone and the adjacent hopper wall. Again, vibration must not be applied unless the material can discharge freely beneath the cone.

These devices have been used very extensively for a wide range of materials, and found to be successful provided the material is not either highly cohesive or highly elastic, as described above. Sizing of such a device presents a challenge, as selecting too small a unit (below the stable rat-hole dimension of the bulk solid) will result in poor performance. Many suppliers of these units will size the unit in relation to the diameter of the hopper, which is not strictly correct as an approach but usually meets with success for the simple reason that their recommendations often result in the specification of quite large devices; not uncommonly two or three metres across, which is above the maximum stable rat-hole diameter of many bulk solids.

In general, the best advice with these devices is to accept the supplier's advice if they can show they have used one of the proposed size (or smaller) with the same bulk solid in the past, but be sure that the bulk solid is truly the same. If this cannot be shown, then it is better not to purchase the device until the supplier has proved its effectiveness with a sample of the bulk solid in a test set-up.

Again it is extremely important to ensure that the vibrator control is sequenced so that it cannot be run unless material is being taken away from the space underneath the device.

2.2.2 Internal screens

Some devices are available which consist of screens of perforated plate or similar, mounted parallel to the hopper cone surface on flexible mounts, and driven by an externally-mounted vibrator.

Experience with these devices is much less widespread, however they have been shown to be effectiveness with some materials. If contemplating the use of such a device then a trial with the bulk solid to be handled is an absolute necessity.

2.2.3 Outlet-plane grid devices

There are on the market several devices which are effectively square screens or grids of a "louvre" pattern which fit horizontally across the outlet of a hopper, and vibrate to encourage the bulk solid to flow through the grid into a short vibrating chute beneath (examples are the "Hogan bin discharger" and the "Siletta discharge aid"). These work in effectively a similar way to the internal cone devices, promoting flow across their entire upper surface. Similarly to conical devices, to be effective such a device must be above the maximum rat-hole dimension of the bulk solid. Retro-fits involve cutting the bottom off the existing hopper. Again a trial is recommended to ensure success with the particular bulk solid to be handled by such a device.

3. AERATION DEVICES

These are probably the second most common flow aid device, so again will be addressed in detail. In general, aeration is most effective for finer materials, say under 20 to 40 microns median particle size. However, the presence of larger particles in a commodity which consists principally of fines will not compromise the effectiveness of aeration.

Aeration works in three ways:

(a) At a low level, it can percolate through the bulk solid helping to overcome any partial vacuum tending to hold the particles together when flow stresses in the hopper tend to move them;

(b) At a higher level, it can cause the bulk solid to expand, reducing interparticle forces; and

(c) When it gets between the bulk of the material and the hopper wall, it can help reduce wall friction.

3.1 Low level aeration

The "vacuum-breaking" effect is beneficial in almost any situation where a fine powder is being discharged. Even when a mechanical extractor or a mass flow hopper with a large outlet is used, the introduction of a very small amount of air (say a quarter to a third of the volume flow rate of bulk solid) will help keep the flow consistent in rate and density. This should be introduced at a level perhaps a quarter of the way up the cone. However, this approach will not promote flow where the hopper outlet is less than the maximum stable rat-hole diameter of the bulk solid, so is hardly really a flow aid in the terms of this paper.

3.2 Aeration as an active flow aid

A higher level of aeration applied at the proper location (between the outlet of the hopper and the rat-hole dimension of the bulk solid) can be used to actively promote flow. With bulk solids which fluidise well (low cohesiveness, hard particles and not too fine a particle size, such as fly ash or cement) the exact means of air injection is less critical than the choice of location; even installations as crude as holes drilled in the wall of the vessel, covered with a disc of filter fabric and with compressed air applied, can be very effective as the air will tend to distribute itself through the bulk solid. Such a crude installation is a cheap emergency retro-fit on an existing vessel, whereas in new or properly redesigned installations a more sophisticated approach (as will be described below) is preferable.

With more cohesive materials such as flour, the air tends to make cracks in the material and blow through these without helping promote flow, and such a crude approach does not work. In such cases, aeration pads of sintered metal or plastic, or (better still) flexible fabric are needed to ensure proper distribution of the air. For best effectiveness, these pads need to be mounted all the way down to the outlet of the hopper, and as far up as is required to prevent rat-hole formation. Often, four lines of aeration mounted at 90 degrees to one another in plan view are found to be most effective. On larger installations (perhaps more than a couple of metres across) it is preferable to "zone" the air flow with a timer and solenoid valves to ensure that all pads receive a fair share of the air flow in turn irrespective of how much material is above them and hence the local resistance to air flow.

With the sort of approach described and bulk solids with good fluidisation properties, it is possible to design very large vessels with almost flat cones (fifteen degrees to the horizontal, or less) making very good use of space. However, such installations need very careful design by experts, and are not really in the realm of "flow aids".

3.3 Very fine materials

For very fine materials such as titanium dioxide or similar, aeration cannot easily get in between the particles so the "reduced wall friction" effect is most effective, normally by lining the entire hopper cone with large segmented aeration pads. These aeration pads are best made from a good quality needlefelt fabric (as used for airslides) supported by perforated plate; sintered plastic or metal, or porous ceramics, have been used sometimes but tend to blind with the fine particles whereas fabric flexes and releases the particles. This approach is effective, but expensive not only in terms of equipment but also air consumption and maintenance. It can be retro-fitted, at a cost. For new installations, the use of gravity-discharge mass flow hoppers may be a more cost-effective solution.

3.4 Air blasters/air cannon

These devices are the exception to the general rule that aeration devices only work well with fine powders, since they are often used for coarser materials such a damp minerals. Firing an air cannon releases a small explosion locally to its injector point, dislodging bridged material. They have been used successfully to help move regular bridges of material from specific locations where flow is not reliable. General concensus suggests that they are best applied in situations where a hopper suffers only from local bridging, e.g. near an outlet. They are only really applicable to fairly large vessels where the energy released during firing can dissipate harmlessely, as if confined this energy can cause structural damage to the vessel itself. They can cause compaction of the bulk solid if applied incorrectly especially with fine powders; however their application is not well understood even amongst vendors of these units.

3.5 General comments on aeration devices

For any aeration system, quality of air is of paramount importance. Water in the air, high moisture content which will be taken up by the bulk solid (or for that matter excessively dry air which will dry out the bulk solid) will cause caking of the bulk solid and make the problem worse. Maintenance of water traps etc. is thus critical.

One drawback of using aeration as an active flow promotion device is that it will tend to expand the bulk solid to a low and somewhat variable density, and can lead to variations in discharge rate. Where the absolute value of discharge rate is not terribly important this is not likely to be a problem, but in instances where careful rate control is required, such as when interfacing to a metering feeder, this will cause difficulties in obtaining proper control. (This comment does not apply to the use of aeration at a low level for "vacuum-breaking" as first described, which will help keep the bulk solid to a more consistent density.

A final point to bear in mind is that aeration (with the exception of blasters) is likely to fail completely when applied to wet, very sticky or coarse materials. In fact it can even make flow problems worse by encouraging drying-out or segregation of such materials.

4. MECHANICAL EXTRACTORS OR DISCHARGERS

All the many types of mechanical dischargers have the same objective; to provide a physical gathering and pushing of bulk solid from across a large inlet dimension above, into a small outlet dimension beneath. This promotes flow very well across the inlet area of the device, and provided that inlet is larger than the maximum rat-hole dimension of the bulk solid, reliable flow is assured.

With mechanical devices, large inlet dimensions can be achieved quite easily, albeit at a cost, so that even very cohesive bulk solids can be handled. In fact it is not uncommon to make the mechanical extractor the full diameter of the vessel so that no convergence is needed on the walls, and such an approach has the added benefit of giving effectively a mass flow discharge pattern in the vessel, thus promoting first-in-first-out storage, and eliminating long-term resident material which can go hard. This is practically a foolproof approach, as the bulk solid cannot hang up anywhere and will always discharge provided the extractor has the power to dig it out. It should be said that this extreme approach is not always required, and some systems work with mechanical extractors of more modest dimensions interfaced to core flow hoppers.

In either event, the point is that these devices can deal with highly cohesive materials which could not be moved by vibration, aeration or other less costly means. In general, they should only be considered where the flowability of the bulk solid is so poor as to make other options (including design for mass flow gravity discharge) impossible or uneconomic.

4.1 Sweep augers

These work by continuously undermining the inventory in the vessel ahead of the screw, and can cope with most bulk solids including damp materials unless they are either so sticky as to clog the screw flighting, containing long fibres which can wrap themselves around the screw, or lumpy and abrasive which would cause high wear.

They are subject to high forces so are strongly constructed with high-torque drives, hence are very expensive. One of the biggest potential problems with such devices is that they produce a rotating stress-field within the silo above, which can very easily destroy the silo unless it has been specially strengthened to an appropriate degree. This makes them almost impossible to retro-fit. Screw breakages occasionally are a possibility especially if the screw hits some highly caked material, so it is worthwhile considering the pattern which allow for replacement of the screw without emptying the silo first, in a large installation.

A variation on the sweep augur which can be retro-fitted to an existing vessel is the "circular bin discharger", with an arch-breaker arm inside the cone of the silo driven from a universal joint at the outlet. The length of the arm is almost equal to the slope length of the hopper cone, and rotates slowly, being free to move and undermine the bulk solid. These devices are still expensive, but sometimes can be applied to an existing silo without the need to cut the whole bottom off the vessel. Again they can discharge materials for which gravity-discharge hoppers would be impossible. If intending to use such a device, a trial is essential.

4.2 Ploughs

Plough dischargers are not commonly used in the UK; although they offer useful advantages, their relatively poor availability has limited the experience available.

4.3 Walking floor and sliding frame dischargers

The "walking floor" discharger consists of a flat silo floor divided into several strips, which oscillate alternately to and fro alongside each other. At one end of the assembly there is a "gap" down which the bulk solid falls when the moving strip retracts, from which it is fed by a screw. The "sliding frame" device again consists of a flat floor, but this time static with a frame sliding backwards and forwards on top of it and encouraging the bulk solid to fall into a slot across the middle of the floor, where again it is taken away by a screw. In each case the discharger mechanism is invariably the full size of the silo plan so there is no converging section.

This type of discharger was mainly developed to deal with bulk solids of the most severe flow characteristics, such as high water content, compressible, highly cohesive materials (e.g. partially dewatered sewage sludge, paper pulp waste etc.) which cannot be handled by any other means. As such they are capable of dealing effectively with these materials, but the equipment cost is very high. These should be considered only when dealing with the most extreme cases of poor flowability.

5. MAKING BETTER USE OF GRAVITY

Gravity is very reliable in itself - it never breaks down or needs maintenance. However, the amount of work available from gravity is strictly limited according to the height available. That means that for bulk solids which are cohesive, and which hence require a significant work input to deform them in flowing down a hopper cone, some care is needed to ensure that gravity alone can overcome the strength of the material.

With relatively free flowing materials, the short height of a cone on a core flow hopper yields enough work to make the material flow reliably. However for materials which have significant internal strength, the outlet size needed for reliable gravity core flow usually becomes unacceptably large (usually half a metre or more, often over a metre). In addition, the last material to be discharged in core flow will have been in static residence since the hopper was first filled (first-in-last-out), so if the material has any tendency to gain strength with time, this last material out will have become very strong in the intervening period and thus very hard to discharge.

5.1 Mass flow

It is in these two respects that mass flow comes into its own:-

(a) For any given material strength, that material will flow out of a much smaller outlet in mass flow than in core flow; and

(b) In mass flow, one has only to take a small amount of material away from the outlet and the whole contents is disturbed, preventing long-term gain in strength and spoilage.

Mass flow requires steeper, smoother walls in the cone of the hopper than does core flow. Just how steep, and what internal surface is best for the come, depends on the frictional characteristic between bulk solid and hopper wall, so it is imperative to have flow tests undertaken with the bulk solid to determine the hopper geometry needed. In the absence of such flow tests, it is impossible to embark on the design of a suitable hopper, however once the flow characteristics are known the actual design procedure for mass flow is extremely reliable, and it is without doubt the most certain way to a reliable hopper.

There are sometimes other considerations which would lead to a preference for mass flow discharge anyway, principally:-

(a) A desire to avoid segregation of fine from coarse (or blend components) in a free flowing material;

(b) A desire for first-in-first-out ("FIFO") discharge for process reasons (e.g. to avoid ageing or cooling, or enable proper stock rotation in foodstuffs); or

(c) Better control of very fine powders (to avoid flushing of aerated material through the bulk from the top surface)

However, setting those aside and looking at the problem purely from the point of view of reliable discharge, mass flow really comes into its own when dealing with *materials which have a significant gain in strength with time.* By disturbing the whole contents every time some material is discharged, and giving a "FIFO" discharge, such materials are handled far more effectively than any discharge aid on a core flow hopper ever could. In fact with highly time-dependent materials, it is probably futile to spend any significant amount of time trying out discharge aids. A decision to convert to mass flow will probably be the most cost-effective solution in the long term, even though it usually means cutting the cone off the hopper and fitting a new one, together with losing some vessel capacity to give the steeper cone walls needed.

There are two principal criticisms often levelled at mass flow:-

(i). The cone section will be taller than for core flow, meaning a reduction in capacity or increase in height; and

(ii). Mass flow hoppers are designed for a particular bulk solid and if the bulk solid changes, the hopper may not deliver mass flow any longer.

In practice, both of these can be overcome very effectively using "plane flow" (wedge-shaped) converging sections which do not require to be so steep, and are more tolerant of changes in bulk solid flow properties. One issue which is critical to obtaining mass flow is the interfacing the feeder at the bottom of the cone; this must allow flow to take place right across the outlet area, which requires a degree of care in selection and interfacing of the feeder. More details of all these issues, together with a detailed analysis of mass flow, can be found in reference (1).

It will of course be appreciated that converting an existing vessel to mass flow can be an expensive exercise (much more costly than fitting discharge aids) although engineering it into a new vessel is relatively cheap, and it is a very low maintenance option - for this reason it is especially worth looking at for new vessels, particularly for highly time-dependent materials as described above.

The ultimate limitation on mass flow is when the bulk solid is so poorly flowing as to require wall slopes near to vertical; in such cases a full-live-bottom with a mechanical discharger the width of the silo is indicated, as described in the section on mechanical devices.

5.2 "Low-friction" linings

Another way of giving gravity a helping hand is to reduce the amount of work it has to do on the bulk solid during discharge, by reducing the friction between the bulk solid and the hopper wall. Where a hopper is "nearly there" in giving mostly quite reliable flow, this can sometimes be an option, however the choice of lining material is not as straightforward as some vendors of linings would make out.

To take some examples, ultra-high-molecular-weight polyethylene's can be very good for some bulk solids (especially damp or wet ones) but bad for others (hard, angular ones). However, the precise grade of polyethylene can make a difference and the one best for some bulk solids will not be the best for others. In the same way, stainless steel can be very good for some bulk solids (not because it is inherently smoother than carbon steel, but because it doesn't rust and become rough when no flow is occurring) - however, choosing the grade of surface finish is extremely important as a mill finish (No. 1 finish) stainless is rougher than a mill finish carbon steel!

The important point is that to choose a "low friction lining" sensibly, some friction tests with the bulk solid against some different lining options are required. Fortunately the cost of wall friction testing alone is trivial compared with the cost of lining even a modest sized hopper, making it an economic exercise.

A word of warning is in order at this point however, regarding the limits of low friction linings; as stated above, they are at their best with "nearly there" hoppers. If a hopper is nearly mass flow, reducing friction will make it mass flow, likewise if it is firmly in the core flow region but almost self-drains then the lining will aid discharge of the last part of the contents. Remember, however, that installing a lining will reduce the transmission of vibration from the outer wall, thus reducing the effectiveness of any external vibrators.

5.3 Inserts

It seems strange to suggest that putting something in the way of the flow channel can actually improve flow, yet in some cases this can be so. The simplest insert is a plain "Chinese hat" suspended in the hopper cone above the outlet, reducing the consolidation pressure on the bulk solid near the outlet and hence making it more free-flowing. More sophisticated is a cone within the hopper cone and the same way up, but with walls somewhat steeper; this can promote mass flow in cones which otherwise would not mass flow. Other insert shapes have been used and shown to be useful.

However the reality of using inserts is that whilst they are often useful to improve the flow pattern in a bin discharging a fairly free-flowing material, they are of limited use with cohesive materials. Both academic research, and practical experience, have a long way to go before the use of inserts can be recognised as anything like a reliable solution to problems of unreliable discharge.

6. CHANGING THE MATERIAL

It is surprising how many cases actually offer the possibility of changing the flow properties of the bulk solid to ease a discharge problem in a hopper. In the food industries it is very common to add a "flow additive" to a product for retail sale in order to render it more free flowing, thus giving the consumer an easier material to handle and also easing problems in packaging machinery etc. Salt, grated cheese, flavouring salts etc. are often treated in this way and although the additives are quite wide-ranging, many act in the same way - to provide a coating of small, hard particles on the surfaces of the softer main particles, preventing the main particles from touching and sticking together. The levels of addition and distribution of the fine "flow additive" need to be very carefully controlled to get the best effect, and any chemical effects of the contamination caused by the flow additives must be considered.

Another approach is to control the moisture content. For example, sugar is "conditioned" (a slow, gentle but very searching drying process) to a low moisture content to prevent it from caking in storage. A further approach is to granulate the material (by spray drying, pelletising etc.) to reduce its surface area, thereby reducing its cohesive strength.

All these approaches are known technology, however their utility is usually dictated by the context. For example, if supplying a poorly-flowing powder of high value to many customers who then have trouble in handling it, it will be economic for the producer to improve the flow characteristics of the powder. Likewise if reclaiming a blend of low-value mineral from a coarse stockpile, mixed with fines from screening together with ultra-fines settled from a lagoon, it may be worthwhile dumping the ultra-fines instead of reclaiming them, to make the balance of the reclaimed material easier to handle. By contrast, if the problem relates to one hopper in a process plant where the bulk solid is quite closely defined, it will probably be better to adjust the hopper to suit the bulk solid.

7. CONCLUSIONS

From the above it will be clear that the choice of approach to improving flow in hoppers is very much dictated by the circumstances. To sum up, the following table may be useful to indicate what approaches may be applicable in which circumstances:-

Generic approach	Specific types	Applicability
VIBRATORS		No good for very fine, wet or very sticky materials.
	External bolt-on	Good for hoppers which "nearly work", easy to retro-fit. Not for severe flow problems
	Internal cone, screen or similar	More effective (& expensive) than external, especially for retro-fit.
AERATION		Works well with fine materials; not for coarse, very sticky or wet materials
	Local (pads)	Useful for many materials except extremely fine ones. Easy retro-fit
	Large area	Good for very fine materials e.g. pigments
	Blasters	For breaking specific areas of bridging; effects somewhat uncontrolled. Retro-fit easy but unpredictable. Danger of structural damage to vessel
MECHANICAL		Can give very large outlet dimensions or "full live bottoms". High maintenance
	Sweep augers and related types	Can handle a wide range of materials. Expensive, but some types can be retro-fitted more economically than others
	Walking floor or sliding frame	For the worst of the worst materials - even more expensive whether new or retro-fit!
HELPING GRAVITY	Mass flow	Especially good for highly time-dependent materials. Very reliable design criteria, expensive to retro-fit but economic for new installations
	Low friction linings	Useful for silos which "nearly work". Economic to retro-fit but critically important to choose the best lining material!
	Inserts	Few types well understood, so hard to plan; low maintenance
CHANGING THE MATERIAL		Reduces handling problems in customers' premises; can be expensive
	Adding "flow aid"	Contamination considerations
	Controlling moisture	Can have other benefits (improving keeping qualities) or costs (reducing weight)
	Excluding fines	Can it be done?
	Increasing particle size	Economic and process considerations

The Institution of Mechanical Engineers is a leading forum for the exchange of knowledge and expertise in the field of mechanical engineering.

A wide range of events is organized by the IMechE, to which all are welcome to attend. For further information about IMechE Conferences, Seminars, Workshops, and other events please contact us for further information.

Visit our website	www.imeche.org.uk
Telephone	+44 (0) 171 222 7899
Fax	+44 (0) 171 222 4557
Or write to	Conferences and Events Institution of Mechanical Engineers 1, Birdcage Walk London SW1H 9JJ

We look forward to seeing you at our future events.